Twenty Lectures on Thermodynamics

Twenty Lectures on Thermodynamics

H. A. Buchdahl
Australian National University

Pergamon Press

Pergamon Press (Australia) Pty Limited, 19a Boundary Street, Rushcutters Bay, NSW 2011
Pergamon Press Ltd, Headington Hill Hall, Oxford OX3 OBW
Pergamon Press Inc., Fairview Park, Elmsford, N.Y., 10523

Cover design by Allan Hondow
Typeset in Brisbane by Filmset Centre Pty Ltd
Printed in Hong Kong by Wing King Tong Co Ltd

Buchdahl, Hans Adolph, 1919-
Twenty lectures on thermodynamics/by H. A. Buchdahl.
Rushcutters Bay, N.S.W.: Pergamon Press (Australia), 1975.
Index.
ISBN 0 08 018299 2
ISBN 0 08 018951 2 Paperback
1. Thermodynamics—Text-books. I. Title.
536.7

CONTENTS

PREFACE

This course of lectures, parts of which I have given at various times over the last few years, presents a coherent, bird's-eye view of phenomenological and statistical thermodynamics, taken strictly in that order. It is largely elementary in character, its main purpose being pedagogic. I have therefore proceeded in a way which here and there may seem idiosyncratic at first sight, even heretical, to the extent that I have constantly allowed physical intuition to take precedence over mathematical niceties, so running counter to the prevailing fashion which tends towards ever greater 'purity' and abstraction. It seems to me that whereas axiomatic developments may well be of great value to the already initiated—though the history of physics does not encourage one to believe that they greatly further its progress—they are likely to prove bewildering for the beginner. What he needs is to be guided as quickly as possible to an understanding of the essential core of the subject. In the case of thermodynamics this is surely the fact of the existence of irreversible processes and the concomitant notion of entropy. Accordingly a discourse on the Second Law and its generic consequences precedes any reference to the First and Zeroth Laws. It develops, moreover, without any formal equations having to be written down; nor do any such equations occur in the introductory material which precedes it. Elsewhere, too, the usual order of development has not been preserved. For instance, the phase rule becomes part of the formal structure of the basic theory, and the idea of the representative ensemble occurs prior to any discussion of the notions underlying statistical thermodynamics. The basic formalism of that theory is established subsequently in a thoroughly heuristic spirit, with repeated appeals to sufficient rather than necessary conditions on the one hand and to the demands of simplicity on the other: the need in a certain sense to accommodate the phenomenological laws being regarded as overriding.

Hopefully this pragmatic approach will lead most rapidly to some familiarity with the subject. As a further aid to this end solutions to all problems which appear at the end of each lecture are provided in full, for to be confronted with problems one cannot solve is worse than useless. Finally, on a number of occasions there are finer points, or sometimes mere questions of detail, which I could not bring myself to ignore altogether but the discussion of which would have unnecessarily encumbered the lectures as such and might have destroyed their continuity. These are relegated to a number of ancillary Notes. The material as a whole is largely self-contained and falls easily within the compass of a one-semester course, the Notes possibly forming a suitable basis for tutorial sessions.

I am grateful to Professor Peter Rastall, Dr Mark Andrews and Dr Peter Sands for reading the manuscript and to Ms Norma Chin for typing it so expertly. Finally I wish to express my warm appreciation of the unfailing courtesy and helpfulness extended to me by the staff of Pergamon Press Australia.

Canberra, September 1974. H. A. BUCHDAHL

LECTURE 1

Preliminary Remarks

The principal aim of these lectures is to present, within a severely circumscribed compass, a particular birds-eye view of both phenomenological and statistical thermodynamics. The course proceeds in a simple, though not entirely unsophisticated, way, so that it may, hopefully, help us to establish a relationship of comfortable familiarity with the subject. Even in an introductory account such as this it is only too easy to get bogged down in the intricacies of logical foundations or mathematical rigour, with the result that we might still be left in the end with the feeling that 'we don't know what it is all about'. Therefore, when some apparently intractable obstacle stands in the way of rapid progress we shall cut the Gordian knot with the sword of the physicist's insight and pragmatism rather than by the exercise of sterile axiomatics. At the same time we relegate incidental detail and the discussion of occasional finer points which should perhaps not be entirely ignored to a series of 'Notes' which appear at the end of the text as such.

To achieve our purpose we must surely concentrate our attention from the outset upon certain particular notions which are often felt to be somehow more recondite than those with which we believe ourselves to be more familiar because we have—so to speak—grown up with them. Entropy, especially, is supposed to have this elusive character, presumably because of the way in which it often appears on the scene as a kind of obscure by-product of superficially rather unrealistic and subtle operations with 'heat engines'. No—we must do better than this, for entropy, whose *raison d'être* lies in the existence in nature of irreversible processes, is the heart of thermodynamics, and we must come to grips with it at the earliest possible moment. We shall indeed do so before temperature or energy or heat have been talked about. In this way we can hope to elucidate its meaning without being first encumbered with a good many ideas which are, in a sense, irrelevant to it.

This lecture began with a reference to 'phenomenological and statistical thermodynamics', and a few words need to be said about the meaning of this phrase. In the first place, thermodynamics concerns itself with certain general limitations which govern the behaviour of large-scale physical systems. By this we may safely understand bodies, or 'things', which can be directly seen and manipulated. Examples might be a liquid in some container, an electric motor, a piece of glass, and the like. Depending upon just what features of a given system we want to investigate and what kind of predictions about its future behaviour we wish to make, we speak more precisely of a 'mechanical system', a 'physico-chemical system', and so on, as the case may be; and each time we construct an appropriate theory.

Thermodynamics is the body of theories appropriate to thermodynamic systems, so that we have to discover first what particular features characterize these. Before we do so we must, however, remind ourselves that as far as all quantitative statements within the theory are concerned we are limiting ourselves exclusively to systems in *equilibrium,* i.e. systems which are such that large-scale observations on them do not reveal the occurrence of any changes taking place in the course of time [*Note* 1]. Within this compass there are two generically distinct thermodynamic theories, called phenomenological and statistical thermodynamics respectively. The first, as its name implies, deals exclusively with systems in bulk, that is to say, with their properties as directly revealed by large-scale observation and experiment. In other words, it takes no cognizance of any small-scale, microscopic, structure and indeed pretends that whether a given system is ultimately atomic or continuous in structure is irrelevant. Concomitantly the amount of detailed information about the system which enters into the theory is small. The output of the theory is correspondingly lacking in detail: for example, it leads to general relations between phenomenological parameters such as specific heats, compressibilities and the like, without being able to say why these parameters have just the particular values which they in fact have.

The statistical theory on the other hand regards the system from the outset as made up of a very large number of atomic constituents ('particles', composite or otherwise) and is based on the premiss that the large-scale behaviour of the system is merely an outward manifestation of the laws governing the motions of the individual particles. To determine these motions in all detail is impossible as a practical proposition so that one has to be content with calculating temporal or other suitable averages over the motions—hence the term 'statistical' thermodynamics. Certain of these averages obey relations which are in all respects formally identical with the relations obeyed by the basic quantities of the phenomenological theory. We thus have, strictly speaking, two distinct but corresponding sets of physical quantities, one phenomenological, one statistical; and the latter constitutes a mechanical model (or realization) of the former. Indeed, one intuitively tends to identify them, but any such identification must, from a conceptual point of view, be treated with caution. At any rate, since much more information is put into the statistical theory—we have to say in detail how the particles are constituted and what mutual forces operate between them—its output is correspondingly rich. In particular it yields the actual values of the phenomenological parameters which previously had to be obtained experimentally.

In the course of the following lectures we shall at first deal exclusively with the phenomenological theory. Its laws are few, very well established and easily understood; and they represent very general, overriding limitations upon the kind of behaviour which thermodynamic systems can exhibit. Subsequently we shall set out to construct the statistical theory in such a way that it must of necessity reproduce the general relations of the phenomenological theory, for I take the view that any theory which does not at least achieve this end is doomed from the outset and must be abandoned. Of course, consistency does not guarantee the correctness of the statistical theory: that can only be established by experiment.

Problems

1.1 In both phenomenological and statistical thermodynamics there occurs a physical quantity often simply called 'entropy', there being an implication that they are somehow identical. Why is this unsatisfactory in principle?

1.2 A well-known book on axiomatic thermodynamics in effect takes a system to be in equilibrium if any change which occurs in it and which affects it alone is such that after it has been restored to its initial condition some overall changes in its surroundings must necessarily have occurred. Would you regard this definition as more satisfactory than the one adopted here?

LECTURE 2

Thermodynamic Systems

What are we to understand here by a 'thermodynamic system'? To answer this question at last we wisely remind ourselves first of what we mean by a 'mechanical system', bearing in mind that we are not interested in the dynamical description of any changes which it may undergo. As a preliminary step, let us specifically contemplate a set of particles mutually joined by straight elastic rods. The whole arrangement is at rest on a table before us, those particles which are in contact with the table being supposed immovable. To describe it in detail we need to give the values of a (finite) set of mutually independent variables, $x_1, x_2, \ldots, x_n$, say, for example the coordinates of some of the particles referred to Cartesian axes, together with the lengths of some of the rods. Any such set of values is a *configuration* of the system. To say that the system has undergone some change from one (equilibrium) configuration to another is merely to say that it has changed from one configuration $x_1' \ldots, x_n'$ to another configuration $x_1'', \ldots, x_n''$. Any such change, then, consists in the displacement of parts of the system relative to each other, that is to say, it is a deformation of the system. Indeed, the fact that every change of any one of the variables $x_1, \ldots, x_n$ describes a certain deformation of the system is sufficiently important to attach a special name to variables which have this character: we call them *d-coordinates*.

Now, two identically constructed systems of the kind just described whose configurations happen to be the same are to all intents and purposes indistinguishable. It is precisely this property which characterizes the class of mechanical systems. In short, a system is *mechanical* if all relevant information about its equilibria is contained merely in the values of a set of d-coordinates.

By way of contrast, we go on to consider a very simple example of a (phenomenological) system—let us always refer to it as our *sample system*—which will turn out to be non-mechanical. It consists merely of a gas. Of course this has to be enclosed by a container; and since we want the volume of this to be variable we may simply take it to be a cylinder, permanently closed at one end and by a movable piston at the other. Further, there shall be some kind of stirrer inside the cylinder by means of which the gas can be agitated from without. Nevertheless, the gas alone constitutes 'the system'; that is to say, neither the enclosure nor the stirrer are to be regarded as part of it.

Now, this system too can be 'deformed', namely by moving the piston [*Note* 2]. A d-coordinate is therefore naturally associated with it, the value of which specifies the position of the piston. We denote it indiscriminately by x; in particular it might, of course, be chosen to be the volume V occupied by the gas. No other non-redundant d-coordinate can be defined. Nevertheless even the most elementary observation shows that the equilibria of this system are not adequately described by the values of x alone. Indeed, we soon find that upon touching the system we may burn our fingers on one occasion and not do so on another. Clearly at least one other variable which is not a d-coordinate must be introduced. I shall call any such variable an *h-coordinate.* In the case of our sample system only one of them is required, and it might, for instance, be taken to be the pressure P of the gas, but this choice is not mandatory.

The state of affairs just described leads us quite generally to the understanding that a thermodynamic system is a system whose equilibria are specified by the values of a finite number n of coordinates (i.e. variables) at least one of which is an h-coordinate. Although this definition appears to be very general, it does imply restrictions on the kind of physical systems which we admit for consideration. In particular, no substances can be present whose properties at a given time depend upon their previous histories. There is an infinity of such histories, whereas we are restricted to a finite number of coordinates.

To keep everything as simple as possible, our main purpose being to gain insight into the subject, we shall very often impose a number of further restrictions, some or all of which are often implicitly taken for granted. The most important of these is that of the n coordinates exactly one is an h-coordinate. We further require that the effects of surface tension can be neglected and that likewise the effects of mutual long-range interactions between parts of the system can be neglected. The first of these assumptions is always satisfied if only the system is large enough; the second would not be justified in the context of a system so massive that the mutual gravitational attraction between its parts is comparable with other forces operating within the system. Finally we require that when any one of the d-coordinates, x_k say, changes slowly enough by an amount δx_k, sufficiently small in absolute value, then the work δW done by the system on its surroundings shall be proportional to δx_k and the ratio P_k of δW to δx_k shall be a function of the coordinates alone. The functions $P_1, \ldots, P_{n-1}$ are called the *(generalized) forces* of the system.

A system which satisfies the various conditions just laid down will be called a *standard system.* The large majority of systems considered in more or less elementary treatments of thermodynamics are of this kind, a fact which is sometimes not sufficiently emphasized. Our sample system is certainly a standard system with two coordinates. At times it is helpful to have a model of a standard system with n coordinates. Somewhat naïvely perhaps, we might take this to consist of $n - 1$ sample systems, any one of the cylinders (taken to be metallic) being in contact with at least one of the others. Each sample system contributes one d-coordinate, so that there are $n - 1$ of these in all. In addition we therefore need just one h-coordinate and this we may, for instance, choose to be the pressure P in any one of the cylinders. Occasionally the theory of a particular non-standard system can be reduced to that of standard systems. The possibility of being able to do so hinges on which particular defining condition of standard systems is not satisfied. It is, for example, not difficult to allow for the presence of surface tension when this is not negligible.

It remains to point out a small sin of omission. When referring to 'a gas' I have throughout said nothing about the nature of this: it might well be a mixture of several gases, capable of reacting chemically with each other. In such a situation the chemist would of course be vitally interested in the proportions in which the various gases are present under conditions of equilibrium. However, the additional variables which must be introduced to describe the internal, i.e. physico-chemical, constitution of a given system do not enter directly into the description of the interaction of the system with its surroundings. It is therefore preferable to deal with the required generalization of the theory after its general foundations have been laid.

Problems

2.1 If the system is just a gas, can you think of an h-coordinate other than *P*?

2.2 A system consists of a block of iron on which work can be done not merely by compressing it but also by the action of an external magnetic field. Is the block a standard system?

2.3 Which, if any, of the following are standard systems: (i) the moon, (ii) a solution of rock salt in water, (iii) a gas consisting of oxygen and hydrogen, (iv) a rain drop? Discuss (iii) in detail.

LECTURE 3

States and Transitions. Adiabatic Isolation. Irreversibility

We are already familiar with the meaning of the two kinds of coordinates of a thermodynamic system K. In future we shall call any particular set of values of these a *state* $\mathfrak{S}$ of K; for which reason the coordinates themselves are often also called 'variables of state'. Of course, unless equilibrium obtains, h-coordinates are not defined, in the sense that experiment does not yield unique values which might be assigned to them. (For example, there is no single well-defined 'pressure' which might be assigned to a gas in turbulent motion.) Accordingly it is meaningless to speak of a 'non-equilibrium state': a system not in equilibrium is simply in no state at all [*Note* 3]. Concomitantly, since all quantitative statements of classical thermodynamics are about states, any predictions about the outcome of experiments involving non-equilibrium processes can be at best qualitative.

A change of any system K from one state to another will be called a *transition* of K. In the course of a transition K may or may not be in equilibrium. When it is, i.e. when K goes through a continuous sequence of states, the transition is called *pseudo-static*, a terminology which reflects the fact that it must proceed 'at an infinitesimal rate' [*Note* 4], since otherwise the appearance of elastic waves, turbulence, and the like would run counter to the equilibrium which is supposed to obtain. If, further, in the course of a pseudo-static transition work is done by the forces P_k alone it is called *quasi-static*. (For example, transitions of a sample system brought about solely by turning the stirrer or alternatively solely by moving the piston, either process occurring at an infinitesimal rate, are pseudo-static and quasi-static, respectively.) Finally, a transition such that K is not in equilibrium in the course of at least a part of it is called *non-static*.

Special conditions are frequently imposed upon transitions which restrict their generality. For example, when all d-coordinates of K are kept fixed by prescription the transition is *isometric*; if we merely require their final values to be the same as their initial values we call it *weakly isometric*. Further, a transition is *infinitesimal* if the final state, and in the case of a quasi-static transition every intermediate state, is sufficiently close to, i.e. 'lies in the *neighbourhood* of' the initial state [*Note* 5]. Again, a most prominent position is occupied in the theory by the class of so-called adiabatic transitions. To define these we must first clearly understand what is meant by adiabatic isolation.

To this end, let us first think about the homely example of an aluminium container filled with water and ice cubes. Upon placing it in a refrigerator we may find that after some time the water has also turned into ice. On the other hand, had we placed the container over the flame of a gas burner the ice cubes would have melted, the evaporation of some or all of the water quite apart. Now contrast this behaviour with what we observe when the aluminium container is replaced by a high-quality thermos flask. True, the ultimate behaviour of the water-ice mixture is the same, yet an altogether different time scale is involved. For instance, upon placing the flask in the refrigerator we may have to wait for days for the water to freeze. It is therefore not unreasonable to suppose that better and better 'thermos flasks' could be constructed until we end up with one which is such that it is to all intents and purposes altogether impossible to melt the ice or freeze the water contained within it merely by placing it into appropriate surroundings [*Note* 6]. The only way to melt the ice is by some mechanical process, say by shaking or stirring.

The particular example just discussed leads us to the following general definition: an enclosure is called *adiabatic* if the equilibrium of a system contained within it can only be disturbed by mechanical means. An enclosure which is not adiabatic is called *diathermic*. Any doubt as to what constitutes 'mechanical means' is removed by prescription; examples are shaking, stirring, the movement of a piston (or more generally any deformation of the system), and the passage of an electric current.

A system which is contained within an adiabatic enclosure is *adiabatically isolated*, and we represent it by the symbol K_0. The transitions of K_0 are *adiabatic transitions*, whether quasi-static or not. Here we should be clearly aware of the distinction between isolation on the one hand and adiabatic isolation on the other. The latter condition is the weaker of the two since it does not forbid arbitrary mechanical interaction between the system and its surroundings. In other words, a system can certainly do work on its surroundings in the course of an adiabatic transition. (We note in passing that the d-coordinates of an adiabatically isolated system can still be varied at will.)

There are evidently two kinds of possible interactions between a system and its surroundings. When it is adiabatically isolated the interaction must be mechanical, and it necessarily involves large-scale, observable motions in the surroundings, such as the displacement of levers, pistons and the like [*Note* 7]. In the absence of adiabatic isolation the interaction may, however, be partly or wholly non-mechanical—synonymously: *diathermic* or *thermal*—and no such observable, large-scale motions need occur.

It remains to talk about another division of transitions into two broad classes, the distinction between which lies at the root of thermodynamics. To this end we must first ask ourselves what one might understand by the 'reversal' of a transition. We contemplate in the first instance only adiabatic transitions. Then the reversal of a transition of K_0 from a state $\mathfrak{S}$ to a state $\mathfrak{S}'$ simply means a subsequent transition which restores the system (adiabatically) to its initial state $\mathfrak{S}$. This may turn out to be in fact impossible; in that case the original transition is called *irreversible*, otherwise it is *reversible*. In the absence of adiabatic isolation, reversal can clearly no longer mean the mere restoration of K to its initial state since this can always be achieved by allowing it to interact suitably with its surroundings. Evidently an additional condition must be satisfied. It is this: that after the restoration of K to its initial state no overall changes in the surroundings remain as the result of the process as a whole. Of course it is not *a priori* obvious that there are any irreversible transitions of an arbitrarily selected system at all. That there are is, in effect, one of the central laws of nature which will occupy us a good deal later on, especially in the next lecture.

Although quasi-static transitions of a standard system K are reversible it is not at all obvious that all reversible transitions of K must be quasi-static. It is in fact an experimental result that they are so in the overwhelming majority of cases. There are, however, 'abnormal systems' which behave differently in this and other respects. We shall return to these at the end of the course.

Problems

3.1 Describe what manifest differences there are between the outcomes of the following alternative transitions of our sample system, i.e. of a gas contained in an adiabatically enclosed cylinder supplied with a piston in the usual way: (i) the piston is moved to and fro very slowly at constant speed so that the volume of the gas is first doubled and then reduced to its initial value; (ii) the volume is first doubled by sudden withdrawal of the piston and then restored to its initial value. (The surroundings of K_0 may be thought of as some external mechanical device.)

3.2 In the case of the system of Problem **2.2,** leaving hysteresis aside as irrelevant here, do you consider the interaction via the magnetic field to be mechanical or thermal?

3.3 A system consisting of a gas contained in an enclosure is being stirred by means of a clockwork mechanism powered by a spring. What is the number of coordinates whose values constitute a state at any given time?

3.4 Can a purely mechanical system undergo 'irreversible changes'?

3.5 A system K consists of two gases each contained in an adiabatic enclosure, the two enclosures being in mutual contact. K is not a standard system. Why not?

LECTURE 4

The Second Law, Entropy and The Entropy Principle

We regard the irreversibility of a given transition of a system as being characteristic of the behaviour of the system as such, not of its surroundings. To this extent the character of its interactions with the surroundings is irrelevant. Both as a matter of principle and to aid our understanding and insight we therefore temporarily contemplate only the simplest kind, namely those which are mechanical. In short, we confine our attention to adiabatic transitions.

Let us consider again our sample system and suppose that the surroundings do work on this system K_0 for a certain time by means of the stirrer. As a result the pressure of the gas will have increased—this is merely a matter of observation—whilst the volume has remained constant. To reverse the transition would require the reduction of the pressure to its initial value, the volume than also having its initial value. How is this to be done? It is useless to turn the stirrer since irrespectively of whether we turn it this way or that, fast or slowly, more work will be done on K_0 and the pressure will increase further. We may, on the other hand, seek to reduce it by withdrawing the piston but then we end up with an increased volume. Upon restoring it to its original value by pushing the piston in we regain either the intermediate value of the pressure, if the to-and-fro process proceeded at an infinitesimal rate, or an even greater pressure if it did not [cf. Problem **3.1**]. In short, the original state of K_0 cannot be restored, i.e. the transition under discussion is irreversible.

We can shift the emphasis a little by rephrasing this conclusion as follows. Given some initial state $\mathfrak{S}$ there is a whole class of states of K_0 such that no transition from $\mathfrak{S}$ to any one of these states is possible. (In particular, if $\mathfrak{S}$ is the state V, P then certainly all those states V', P' such that $V' = V$, $P' < P$ belong to the class of states in question, no matter how small $|P' - P|$ may be.) We can sum up the whole content of the conclusion reached so far somewhat formally as follows: in any neighbourhood of a given state $\mathfrak{S}$ of K_0 there are states $\mathfrak{S}'$ inaccessible from $\mathfrak{S}$ [recall *Note* 5]. Moreover, it is obvious in the case of the sample system under discussion that if the state $\mathfrak{S}'$ is adiabatically inaccessible from $\mathfrak{S}$, then $\mathfrak{S}$ is (adiabatically) accessible from $\mathfrak{S}'$. Setting our special system aside now, the results of innumerable experiments confirm that the two main conclusions just stated hold quite generally for (standard) systems of arbitrary complexity [*Note* 8].

Suppose now that $\mathfrak{S}$ and $\mathfrak{S}'$ are arbitrarily selected states of some adiabatically isolated system K_0, such that $\mathfrak{S}$ is inaccessible from $\mathfrak{S}'$ (and therefore $\mathfrak{S}'$ accessible from $\mathfrak{S}$). This specification evidently contains a time order in the following sense: given merely that K_0 was in the state $\mathfrak{S}$ at some time and in the state $\mathfrak{S}'$ at some other time we can say that

$\mathfrak{S}'$ was necessarily the later state. With this idea of an ordering amongst the states of K_0 in mind, let us contemplate some unspecified set of selected states $\mathfrak{S}_1$, $\mathfrak{S}_2$, . . . of K_0. We first represent $\mathfrak{S}_1$ by an arbitrarily situated point p_1 on a straight line. Next $\mathfrak{S}_2$ is represented by a point p_2 of which we merely require that it lie to the right of p_1 if $\mathfrak{S}_1$ is inaccessible from $\mathfrak{S}_2$, to the left of p_1 if $\mathfrak{S}_2$ is inaccessible from $\mathfrak{S}_1$, and that it be coincident with p_1 if $\mathfrak{S}_1$ and $\mathfrak{S}_2$ are mutually accessible. (The three cases respectively correspond to the transition from $\mathfrak{S}_1$ to $\mathfrak{S}_2$ being irreversible, impossible, and reversible.) We continue this construction by representing the states $\mathfrak{S}_3$, $\mathfrak{S}_4$, . . . by points p_3, p_4, . . . in turn in such a way that, if p_j and p_k are any two points, p_k lies to the right of p_j, to the left of p_j, or coincides with p_j according as the transition from $\mathfrak{S}_j$ to $\mathfrak{S}_k$ is irreversible, impossible or reversible, respectively. Finally we attach to each such point a number which can be chosen at will subject to the sole condition that these numbers increase monotonically as we proceed from left to right. (Coincident points of course take the same number.)

Each state is a set of values of the coordinates $x_1, \ldots, x_n$ of the system. The procedure just described therefore amounts to the definition of a certain 'state-function' $s(x_1, \ldots, x_n)$ (or simply $s(\mathfrak{S})$); for this is only another way of saying that a correspondence has been established between the set of states $\mathfrak{S}_1$, $\mathfrak{S}_2$, . . . and a set of numbers s_1, s_2, . . . [*Note* 9]. Any particular $\mathfrak{S}_j$ is labelled by a number $s_j = s(\mathfrak{S}_j)$, this labelling having the fundamental property that a transition from a state $\mathfrak{S}_j$ to a state $\mathfrak{S}_k$ is possible if and only if $s_j \leqslant s_k$ (equality implying reversibility). We call s_j the *empirical entropy* of $\mathfrak{S}_j$ (or of K in the state $\mathfrak{S}_j$), and $s(\mathfrak{S}_j)$ the *empirical entropy function* of K. Evidently if K_0 undergoes a transition from $\mathfrak{S}_j$ to $\mathfrak{S}_k$, the empirical entropy of the final state can never be less than that of the initial state.

Now, however large the set of states $\mathfrak{S}_1$, $\mathfrak{S}_2$, . . . may be—it may even be infinite—it certainly does not contain all possible states in which K_0 may find itself: they form a continuum [*Note* 10]. It is therefore impossible to construct a labelling of the states of K_0 by the simple method adopted before. Yet, the distinction between the continuum of states and a sufficiently large set $\mathfrak{S}_1$, $\mathfrak{S}_2$, . . . selected from amongst them, however important mathematically, is physically vacuous. Being loath to abandon the particular conceptual approach which led so easily to an understanding of the idea of entropy we cut the Gordian knot by simply declaring that such a 'good' function $s(x_1, \ldots, x_n)$ [*Note* 11] with the required properties always exists. In short, we lay down the following law, intended to be a formulation of the *Second Law of Thermodynamics* (so called for historical reasons):

There exists an ordering amongst the states of any thermodynamic system K, reflected in the existence of a function s of the coordinates of K such that if s′, s″ are the values of s for arbitrarily selected states $\mathfrak{S}'$, $\mathfrak{S}''$ *of K, then* $\mathfrak{S}''$ *is adiabatically inaccessible from* $\mathfrak{S}'$ *if and only if* $s'' < s'$.

We call s an *empirical entropy function* of K, as before, and that it is a good function is left understood. It is not uniquely determined, in as far as $\sigma(s)$ is an equally acceptable empirical entropy function if $\sigma(s)$ is any good monotonically increasing function of s. This arbitrariness runs precisely parallel with that which we encountered when we were numbering a set of discrete representative points on a line: the sole condition which these numbers had to satisfy was that they should increase monotonically from left to right. Nonetheless, there are preferred functions σ, but it is perhaps better to deal with these later on. At any rate, from the outset the Second Law in its present form places explicit emphasis upon the existence of irreversible processes. In so doing it directly attaches an intuitively transparent meaning to entropy without the encumbrances of having to take into account any of the other principal laws of thermodynamics and the various quantities, such as energy or heat, whose definitions derive from them.

Finally, according to the Second Law the transition from $\mathfrak{S}'$ to $\mathfrak{S}''$ is possible only if $s'' \geqslant s'$, i.e. *the entropy of the final state of an adiabatic transition is never less than that of the initial state.* This is nothing other than the *Principle of Increase of Entropy* (or 'Entropy Principle' for short), and this appears here to all intents and purposes merely as a rewording of the Second Law.

Problems

4.1 Experiment shows that PV^{γ} is constant during a reversible adiabatic transition of a certain gas (γ is a given constant > 1). What states are adiabatically inaccessible from the state $\mathfrak{S}_1[P_1, V_1]$?

4.2 Is there any substantial reason why in the statement of the Second Law the inequality $s'' < s'$ could not be replaced by $s'' > s'$?

4.3 Strictly speaking our form of the Second Law contains an omission. Can you say what it is, bearing in mind what you know about mechanical systems?

LECTURE 5

The First Law, Energy and Heat

We return briefly to mechanical systems. Given any configuration, we know all about the forces which hold the system in equilibrium, such as gravity, elastic forces in connecting rods, and the like. In a change from one configuration to another, i.e. in the course of deforming the system, a certain amount of work will, in general, have to be done against these forces, and this depends solely on the initial and final configurations [*Note* 12]. We can put this in another way: for any mechanical system there exists a function E of its coordinates such that the work done on the system in any change of configuration is equal to the increase in the value of E. This function, called the energy of the system, is defined only to within an arbitrary additive constant. Evidently if a particular change of configuration is such that the work done happens to be zero, then the initial and final values of E are equal: this is what is meant here by the 'conservation of energy'. In short, the energy of every mechanical system is conserved; but in saying this we must bear in mind that E is solely a function of d-coordinates [*Note* 13].

We now go over to thermodynamic systems. Work is done here, too, when the d-coordinates vary. It is, however, not true that the amount of work in a transition depends solely on the deformation it has undergone, i.e. on the extent to which the d-coordinates have varied. That this is so becomes obvious on reflecting that work can be done on the system simply by stirring it, and in that process the values of the d-coordinates do not change at all. Furthermore there is no reason why, in general, the final values of the h-coordinates should differ from their initial values. As far as the system is concerned, the work done on it has apparently simply 'disappeared'. At first sight it therefore looks as if no energy function could usefully be ascribed to a thermodynamic system. How is this unhappy situation to be rescued?

A possible answer to the preceding question suggests itself as soon as we recall that a thermodynamic system K can interact in two ways with its surroundings, namely mechanically and thermally (i.e. non-mechanically). Once again we are led to exclude thermal interactions; that is to say we consider only adiabatic transitions for the time being. In this situation the results of innumerable experiments show that

the amount of work done by a system in an adiabatic transition depends on the terminal states alone.

This, indeed, is the so-called *First Law of Thermodynamics*. To leave no room for doubt, let it be emphasized that according to this law the work W done does not depend on the manner in which the transition of K_0 between given states proceeds, i.e. whether it is reversible or not, and so, in particular does not depend upon any intermediate states through which K_0 may happen to pass in the course of the transition.

Further progress is now quite straightforward. Not to be hindered by mere appearances, we first simplify the notation a little by writing, in appropriate circumstances, x for the whole set $x_1, \ldots, x_n$, and of course y for $y_1, \ldots, y_n$, and so on. A state $\mathfrak{S}'$ is then a set of values x' of the coordinates x, and this is occasionally made quite explicit by writing $\mathfrak{S}'[x']$ in place of $\mathfrak{S}'$ alone. Now, according to the First Law, the amount of work done by a system in any adiabatic transition from a state $\mathfrak{S}^*[x^*]$ to a state $\mathfrak{S}^{**}[x^{**}]$ is merely a function $F(x^*, x^{**})$ of these states (F depending of course on the structure of the given system). Hence if K_0 undergoes a transition from $\mathfrak{S}'[x']$ to $\mathfrak{S}''[x'']$ and then a subsequent transition from $\mathfrak{S}''[x'']$ to $\mathfrak{S}'''[x''']$ the total amount of work done by it is $F(x', x'') + F(x'', x''')$. However, we know from the outset that in any transition from $\mathfrak{S}'$ to $\mathfrak{S}'''$ the work done by K_0 is $F(x', x''')$, so that the identity

$$F(x', x'') + F(x'', x''') = F(x', x''')$$

must be satisfied for all values x'' of the coordinates of the intermediate state. In other words, in forming the sum on the left the x'' must cancel out and this is only possible if $F(x', x'')$ has the generic form

$$F(x', x'') = U(x') - U(x''),$$

where U is a certain function which is as usual assumed to be a good function. It is called the *internal energy function* of K or, more briefly, its *energy*. It is defined, i.e. can be measured, only to within an arbitrary additive constant. Thus, simply expressed, the First Law declares that in any adiabatic transition the work W_0 done by the system is equal to the decrease $-\triangle U$ of its energy:

$$W_0 + \triangle U = 0.$$

We thus have a situation which is strongly reminiscent of that encountered in the case of mechanical systems.

Now let the condition of adiabatic isolation be abandoned. When K undergoes some transition between given states $\mathfrak{S}'$, $\mathfrak{S}''$ the best we can do is to measure the amount of work W actually done by K_0: there is no reason why it should be just W_0. In other words, the quantity $W + \triangle U$ will in general fail to vanish. It is therefore usual to introduce the abbreviation

$$Q = W + \triangle U,$$

and for historical reasons one rather quaintly calls Q the 'heat absorbed by K' in the transition. However, let us be clear that this *heat* Q is simply the difference between two mechanical quantities, namely the difference between the amount of work done by the system in the actual transition between given states and the amount of work which would have been done in an adiabatic transition between the same states [*Note* 14]. In particular, a transition is adiabatic if and only if no heat is transferred between the system and its surroundings in the course of it. This is of course not a definition: it is merely a somewhat trivial consequence of the First Law, and to state the latter we already had to know what was meant by an adiabatic transition.

We have already supposed that U is a good function [recall *Note* 11]. Apart from this, experiment shows that the energy of any given system has a least value, i.e. it has a lower bound. On the other hand the energy of the vast majority of systems has no upper bound. Unless the contrary is stated explicitly, I shall therefore suppose throughout that the energy of any system can be increased indefinitely, and that this can indeed be done even by means of isometric processes alone.

A 'perpetual motion machine of the first kind' is intended to be some cyclic mechanism, however complex, which can continue indefinitely to do work on its surroundings with which it can interact only mechanically. However, at the end of any number of cycles, being again in its initial state, $\triangle U = 0$ and, since Q also vanishes, $W = 0$. There are therefore no such mechanisms; and the failure of all attempts to construct them is simply evidence of the validity of the First Law, or, as one sometimes says, of the conservation of energy [*Note* 15].

Finally, a word about the additivity of energy. Suppose we consider two systems K_A and K_B jointly and regard them as constituting a compound system K_C. It does not matter whether K_A and K_B are in mutual contact or not. If they are standard systems the work done by K_C in any transition is simply the sum of the amounts of work done separately by K_A and K_B; for on the one hand there are no surface forces which might have to be taken into account when they do happen to be in mutual contact, whilst on the other, their relative positions are irrelevant because of the absence of mutual long-range interactions. Since energy changes and subsequently heat are defined solely in terms of the amounts of work done on the various systems under the appropriate conditions, it follows at once that $U_C = U_A + U_B$ and $Q_C = Q_A + Q_B$. In short, in this sense energy and heat are additive, but we must not forget that we have only standard systems in mind.

Problems

5.1 On any reasonable time scale the earth will continue indefinitely to revolve about the sun in a nearly elliptic orbit. Is this an example of a perpetual motion machine?

5.2 A rigid adiabatic enclosure is divided into two parts by means of an internal partition. Initially the first compartment contains a gas whilst the second is evacuated. The partition is subsequently removed so that the gas will fill the whole enclosure. What can you say about the value of $\triangle U$?

5.3 For the sake of argument, suppose that a system consists of a fluid contained in a rigid adiabatic enclosure. (The system is artificial in that there are no deformation coordinates. On the other hand, the usual stirrer is to be present.) Determine an empirical entropy function.

LECTURE 6

The Zeroth Law

At the end of the last lecture we already had occasion to regard two standard systems K_A and K_B jointly as constituting a compound system K_{AB}. We do the same here, but now specifically suppose K_A and K_B to be in mutual contact. Depending upon the physical nature of the boundary separating them, K_{AB} may or may not be a standard system. Thus, if the boundary is adiabatic there is nothing to prevent K_B from being in any arbitrarily chosen state $\mathfrak{S}_B$ no matter what the state $\mathfrak{S}_A$ of K_A may happen to be. This implies that K_{AB} has two h-coordinates, so that it is not a standard system. On the other hand, when the boundary is diathermic arbitrarily chosen states of K_A and K_B can no longer coexist. What can we say about the condition or conditions which these states, taken together, must satisfy?

To begin with, let us agree that if K_A and K_B are any two standard systems, the phrase 'K_A is in equilibrium with K_B' shall have the following specific meaning: when K_A and K_B are brought into mutual diathermic contact (their d-coordinates of course being kept fixed) they will continue to be in the states $\mathfrak{S}_A$, $\mathfrak{S}_B$ in which they were prior to the establishment of contact. In other words, K_{AB} will be in a state, no change in the states of K_A and K_B having taken place. We note in passing that the manner in which contact is established is irrelevant, in the sense that it does not matter which parts of K_A and K_B happen to have been brought into mutual contact.

Now, let K_C be a third standard system in some state $\mathfrak{S}_C$. We can contemplate the situation in which it has been found by observation that K_A is in equilibrium with K_C and that K_B is also in equilibrium with K_C. Whether K_A is then in equilibrium with K_B we cannot say: only experiment can tell us that. It turns out that mutual equilibrium between K_A and K_B does in fact necessarily obtain. This general empirical result, then, is contained in what is conventionally called the *Zeroth Law of Thermodynamics:*

if each of two standard systems is in equilibrium with a third standard system then they are in equilibrium with one another.

Beyond this point we can proceed most simply by making use of the freedom we have to choose K_C, say, to be a system of a particularly simple and convenient kind. We take it to be any standard system which has only a single coordinate, say z (of whatever kind) subject solely to the condition that in any adiabatic transition its final value is either always greater than or else always less than its initial value. A simple, fairly realistic example of such a system is that of an amount of gaseous helium within an enclosure of fixed volume. If z be taken to be the pressure P of the gas then adiabatically this can never decrease, whereas if z had been taken to be $1/P$ then it could only decrease.

Next we investigate by experiment those states of K_{AC} in which K_C is in a fixed, arbitrarily chosen state $\mathfrak{S}'_C$; that is to say, we vary the coordinates $x_1, \ldots, x_n$ of K_A in such a way that K_A remains in equilibrium with K_C, z having the fixed value z'. It turns out that every state $\mathfrak{S}_A$ of this kind satisfies just one condition of the form $\tau_A(x_1, \ldots, x_n) = z'$, where τ_A is a good function [*Note* 16]. Likewise, if K_B is a third system, with coordinates $y_1, \ldots, y_m$, the states of K_B in equilibrium with K_C in the state $\mathfrak{S}'_C$ are those which satisfy one condition of the form $\tau_B(y_1, \ldots, y_m) = z'$. Bearing the Zeroth Law in mind, it follows that K_A and K_B are in equilibrium if and only if their states satisfy the condition

$$\tau_A(x_1, \ldots, x_n) = \tau_B(y_1, \ldots, y_m).$$

The functions $\tau_A, \tau_B, \ldots$ of systems $K_A, K_B, \ldots$ are particular *empirical temperature functions* of these systems, and the value t of the empirical temperature function τ of a system K in the state $\mathfrak{S}$ is the *empirical temperature* of K (or of $\mathfrak{S}$) [*Note* 17]. The answer to the question put at the beginning of this lecture can now be given as follows: with any standard system K we can associate a (good) empirical temperature function such that any two such systems are in mutual equilibrium if and only if the values of these functions are equal, i.e. the two systems have the same temperatures.

The functions $\tau_A, \tau_B, \ldots$ are not uniquely defined—that is why, a moment ago, we spoke of 'particular' rather than 'the' empirical temperature functions. In the first place the condition $t_A = t_B$ for the mutual equilibrium of the systems K_A, K_B can equally well be written $\theta(t_A) = \theta(t_B)$, where θ is any good, monotonic function; and then one may take $\hat{\tau}_A = \theta(\tau_A)$, $\hat{\tau}_B = \theta(\tau_B), \ldots$ as the empirical temperature functions and *their* values as the empirical temperatures of the systems in question. This device corresponds to going over from the coordinate z of K_C adopted initially to a new coordinate $\hat{z}$, where $\hat{z} = \theta(z)$, writing the condition of equilibrium between K_A and K_C now as $\hat{\tau}_A(x_1, \ldots, x_n) = \hat{z}'$. Equivalently we can regard $\hat{z}$ as being the one coordinate of another reference system $\hat{K}_C$ altogether, in which case $\hat{z} = \theta(z)$ is the condition of equilibrium between K_C and $\hat{K}_C$.

The arbitrariness inherent in the definition of empirical temperature just described is reflected in practice in the use of a variety of 'temperature scales'—Celsius, Fahrenheit, and the like. That is not to say that later, upon transcending the compass of the Zeroth Law alone, we may not come to prefer some particular scale, i.e. a particular temperature function over all others, just as we may come to prefer some particular entropy function over all others. For the moment we shall do no more than to narrow down all possible choices by a convention based upon the following reflection. Starting from some given state of an adiabatically isolated system, a sequence of isometric transitions leads to greater and greater values of the energy U. At the same time the empirical temperatures will form either an increasing or a decreasing sequence. We are therefore at liberty to choose the empirical temperature function so that t is always an increasing function of U. It is worth noting in passing that the values of the empirical entropy s form an increasing sequence at the same time. In short, we now always have $\partial U/\partial t > 0$ and $\partial s/\partial t > 0$.

The convention which we have just adopted brings with it that of two systems the one which has the higher temperature is also the 'hotter' of the two; and when these systems are brought into mutual diathermic contact, but are otherwise adiabatically isolated, the establishment of equilibrium entails the transfer of heat from the initially hotter to the initially colder system. It is in effect largely for this reason that one wants to know the temperature of this body or that in everyday life. To determine it one uses a *thermometer*. A moment's reflexion shows that such an instrument is precisely a system of the kind of

K_C, i.e. one which has only a single coordinate z. Indeed, the particular example, already described, of a quantity of gaseous helium maintained at constant volume, constitutes a so-called 'constant volume gas thermometer' and it is often used in practice. The condition $\partial U/\partial t > 0$ will then be satisfied provided we take z to be an increasing function of P [*Note* 18].

The set of all states of a given system K which have a common temperature t is an *isothermal* of K. By definition of the empirical temperature function they satisfy the equation $\tau(x_1, \ldots, x_n) = t$, and this, or any equation $\varphi(x_1, \ldots, x_n, t) = 0$ equivalent to it, is usually called the *(thermal) equation of state* of K. It is very often convenient to take t as the h-coordinate of a given standard system. This amounts to eliminating the original h-coordinate by solving the equation $\tau = t$ for it. Since τ is a good function this can always be done.

Problems

6.1 Give an example of a simple non-standard system to which the Zeroth Law is clearly irrelevant.

6.2 Why is the commonly used phrase 'the temperature in the sun was so-and-so' nonsense?

6.3 We are all familiar with the usual kind of mercury-in-glass thermometer. Why is a water-in-glass thermometer not satisfactory in the range 0°C—10°C?

6.4 Fourier's well-known equation is a differential equation whose solutions describe the temperature distribution in a given body subject to appropriate boundary conditions. If we adhered rigorously to the point of view adopted in these lectures we would have to declare this equation to be meaningless. Why?

LECTURE 7

Entropy Revisited: Metrical Entropy and Absolute Temperature

Entropy, regarded as a physical quantity, is a formal concomitant of our recognition of the existence of irreversible processes. We are, indeed, already well aware that it serves to quantify the relative adiabatic accessibility of states, but we have not yet investigated how an empirical entropy function s may in practice be associated with any given system. It is therefore incumbent upon us to rectify this omission, enquiring at the same time whether there is perhaps amongst all possible such functions one which is preferred over the rest. Our position is now stronger than it was in the course of the fourth lecture to the extent that we can now draw also upon the content of the First and Zeroth Laws. The invalidity of one or both of these would not by itself necessarily entail the invalidity of the Second; but the empirical determination of s would be a matter of some difficulty. On the other hand, once the energy function is available, we could, on the simplest level adopt a procedure which, though generally redundant, is worthy of consideration in that it will help to deepen our insight.

To this end, given a system K, select some arbitrary, convenient state $\mathfrak{S}_1$. If $\mathfrak{S}$ is any other state of K we first adjust the d-coordinates reversibly and adiabatically so as to achieve a state $\mathfrak{S}^*$ isometric with $\mathfrak{S}_1$. This is always possible, the consequent variation of the h-coordinate being allowed to take care of itself. In the transition from $\mathfrak{S}$ to $\mathfrak{S}^*$ the entropy has a fixed value. To $\mathfrak{S}^*$, on the other hand, we may, as we already know, assign a value of the entropy which is simply numerically equal to that of the energy: $s^* = U^*$. In short, the value s of the entropy of the state $\mathfrak{S}$ is U^*; and a particular empirical entropy function of K has thus been determined (recall *Note* 18).

We expect the procedure just described to involve a good deal of redundancy since we have made no use of our knowledge concerning the detailed behaviour of K which is contained in the explicit form of the functions P_k. Of course, we had no choice in the matter: all the transitions we contemplated were adiabatic and were necessarily irreversible, disregarding the very special cases in which $\mathfrak{S}^*$ happened to coincide with $\mathfrak{S}$. Conversely, if a detailed quantitative description of transitions is to be possible they must be reversible; but then we cannot in general require them to be adiabatic. We therefore now abandon the condition of adiabatic isolation.

A finite reversible transition is a succession of transitions between neighbouring states [recall *Note* 5]. The energy difference between these is dU, and at the same time we write dQ for the heat transferred to K and dW $(= \sum_{k=1}^{n-1} P_k\, dx_k)$ for the work done by K. We must be clearly aware that whereas dU is the total differential of U, the replacement of Q by

dQ and of W by dW is merely a mnemonic device to indicate that we are dealing with infinitesimal quantities [*Note* 19]. The transcription of the equation $Q = \triangle U + W$ which we used in the fifth lecture is thus $dQ = dU + dW$, or more explicitly

$$dQ = \sum_{k=1}^{n} X_k \, dx_k ,$$

where, by way of abbreviation, $X_k = P_k + \partial U/\partial x_k \ (k \neq n)$, $X_n = \partial U/\partial x_n$. In the course of this infinitesimal transition the value of the empirical entropy changes by the amount

$$ds = \sum_{k=1}^{n} s_k \, dx_k,$$

where $s_k = \partial s/\partial x_k$. Now, an adiabatic (reversible) transition is characterized by the vanishing of dQ for each of its elements. Then, however, ds necessarily also vanishes, and it follows at once that dQ and ds must be proportional to each other [*Note* 20]. In short, once the explicit form of dQ has been written down, we are assured that as a consequence of the Second Law functions $\lambda(x)$ and $s(x)$ can always be found [*Note* 21] so that, identically,

$$dQ = \lambda \, ds.$$

Then, provided an arbitrarily given pair of states $\mathfrak{S}'$, $\mathfrak{S}''$ can be reversibly connected at all, the difference between their (empirical) entropies is $\triangle s = \int \Sigma \, (X_k/\lambda) dx_k$. The path of integration is here, of course, irrelevant since from the outset $\triangle s$ is the difference between two values of a state-function.

We have still made no use of the availability of the empirical temperature function. Since this concerns the joint behaviour of pairs of systems we now have to think about a standard system K_A which together with a standard system K_B makes up a compound standard system K_C. Empirical temperature and entropy functions $t_A(x)$, $t_B(y)$, $s_A(x)$, $s_B(y)$ may be taken to have been determined. Then it will obviously simplify matters if we definitely choose t_A to be the h-coordinate of K_A and t_B that of K_B, for then we can simply write t for both of them since $t_A = t_B$ at all times. It is further most convenient to eliminate, for the time being, one of the $n-1$ d-coordinates of K_A in favour of s_A and likewise one of the $m-1$ d-coordinates of K_B in favour of s_B. Then K_C has $n+m-1$ coordinates, i.e. s_A, s_B, t, together with the remaining $n+m-4$ d-coordinates. Since K_C is a standard system in its own right it has an empirical entropy s_C which is a function of the coordinates just enumerated.

Now, recalling what we learned at the end of the fifth lecture, $dQ_C = dQ_A + dQ_B$, and so

$$ds_C = (\lambda_A/\lambda_C) ds_A + (\lambda_B/\lambda_C) ds_B.$$

We see immediately that s_C is a function of s_A and s_B alone, for were it to depend on any of the other coordinates their differentials would have to appear on the right. Therefore λ_A/λ_C and λ_B/λ_C depend on s_A and s_B only. Then if λ_C depended on any coordinates other than s_A, s_B and t these coordinates would have to occur in both λ_A and λ_B. This, however, cannot be since the coordinates of K_A and K_B other than t do not occur in λ_B and λ_A respectively, so that the coordinates in question could not disappear on forming the ratios λ_A/λ_C and λ_B/λ_C. By the same token t can occur in λ_A, λ_B, λ_C only through a common factor, $T(t)$, say. The other factor of λ_A depends only on s_A and that of λ_B only on s_B, which means that there are functions $S_A(s_A)$ and $S_B(s_B)$ such that

$$dQ_A = T(t)\, dS_A(s_A), \quad dQ_B = T(t)\, dS_B(s_B).$$

Finally, dQ_C is generically $T(t)$ times the total differential of a function S_C of s_A and s_B only, and the results just arrived at show that

$$S_C = S_A + S_B,$$

granted that we ignore an additive constant of integration which has no experimental significance in any event.

The fact that one and the same function $T(t)$ serves as an integrating denominator for the 'element of heat' dQ of all systems whose temperatures are measured on a common scale t [*Note* 22] lends to $T(t)$ a certain universal character, and it is therefore called the *absolute temperature (function)*. Not only has a preferred temperature function thus emerged, but—for any given system—a preferred entropy function as well, namely $S(x)$, whose total differential is dQ/T. S is called the *metrical entropy (function)* of K. It is characterized by its additivity in the sense expressed by the equation $S_C = S_A + S_B$.

The absolute temperature scale is defined only to within a constant factor since we can always write $(a^{-1}T)d(aS)$ in place of TdS, where a is any constant. We agree once and for all that to choose the sign of a so that $d(aS)/ds > 0$, whilst factors $|a|^{-1}$ and $|a|$ may simply be left understood in T and S respectively. As a result we are now assured that the Entropy Principle need not be reworded when it is taken to refer to the metrical entropy. We still have to say something about the actual sign of T, but we conveniently defer consideration of this question to a later lecture.

The additivity of entropy referred to composite standard systems. However, a system K_C made up of two standard systems which are adiabatically isolated from each other is not itself a standard system. In that case the relation $S_C = S_A + S_B$ is to be taken as defining the metrical entropy of K_C, and it is again subject to the Entropy Principle [*Note* 23]. Considering now an arbitrary transition of some system K, $\triangle(S + \bar{S}) \geqslant 0$ where barred quantities refer to the surroundings $\bar{K}$ of K. We hopefully suppose that we can arrange the transition of $\bar{K}$ to be reversible, and then $\triangle\bar{S} = \int d\bar{Q}/\bar{T} = -\int dQ/\bar{T}$. We conclude that $\triangle S \geqslant \int dQ/\bar{T}$. This inequality is often stated with T replacing $\bar{T}$, but then the meaning of T is obscure, to say the least.

Finally, a word about *perpetual motion machines of the second kind*. Such a device is intended to be an adiabatically isolated system whose d-coordinates vary cyclically and so continues to do work on its surroundings at the expense of its own energy. At the end of any number of cycles U, and therefore S, would have decreased, which is impossible [*Note* 24]. The failure of all attempts to construct such perpetual motion machines in fact constitutes an empirical verification of the validity of the Second Law. In any event, the term 'perpetual motion' is altogether a misnomer here, granted that U has a lower bound.

Problems

7.1 Experiment shows that a certain substance has the isothermals $PV = t$ whilst its energy function is $U = (\alpha + \beta \ln V)t$, where α and β are positive constants. Show that the absolute temperature function is given by $T = at^{1/(\beta+1)}$, where a is a constant.

7.2 We appear to have dealt solely with the irreversibility of adiabatic transitions. Are such transitions also irreversible in the wider sense: meaning, as already explained, that after the initial state of K has been restored by means of a subsequent, necessarily non-adiabatic, transition some overall change in the surroundings $\bar{K}$ must remain?

7.3 Can you think of a physical quantity q which obeys a 'one-sided' conservation law, i.e. either $\triangle q \geqslant 0$ always or $\triangle q \leqslant 0$ always? (Mass, obeys a 'two-sided' conservation law: in an isolated system $\triangle m = 0$, always.)

7.4 A classical ideal gas is a substance which has the equation of state $PV = RT$ ($R =$ constant). Determine its metrical entropy function $S(V, T)$.

LECTURE 8

Constitutive Coordinates. Characteristic Functions

Already at the end of the first lecture we took note of the eventual need to enlarge our formalism so that we might be able to describe the detailed physico-chemical constitution of a given system K. In general this will consist of a number φ of distinct physically homogeneous parts. Each such part, called a *phase*, has visibly a definite boundary. At the same time K is made up of a number z of chemically distinct substances, and each of these is called a *constituent* of K. To make a complete description of the equilibrium of K possible, we now adjoin to the familiar *n external* coordinates x_k the φz *constitutive* (or *internal*) coordinates n_{aj}, where the *mole-number* n_{aj} is the number of moles of the jth constituent C_j present in the *ath* phase. A set of values of the x_k is now an *external state*, a set of values of the n_{aj} is an *internal state* and the two jointly make up a *state* of K.

To understand what is at stake we first contemplate a system K consisting of a single chemically inert phase. Were this system *closed*—like all others encountered so far—that is to say, were it contained in an enclosure impermeable to matter, the n_i ($\equiv n_{1i}$) would of course be fixed. This now artificial situation is relieved by taking the phase to be an *open* system, meaning that it can exchange matter with its surroundings. To have such an exchange firmly under our control, we may think of an amount dn_j of C_j to be added reversibly to the phase through a tube leading into its interior, this tube containing a semi-permeable membrane which permits the passage of the particular constituent C_j alone. This addition, supposedly taking place whilst all other d-coordinates remain fixed, requires a certain amount of work, proportional to dn_j, to be done on K: let it be $\mu_j dn_j$, where μ_j, the *chemical potential* of C_j, is a function of all the coordinates. Then in a general infinitesimal reversible transition of the open phase the work done by it is

$$dW = \sum_{k=1}^{n-1} P_k\, dx_k - \sum_{j=1}^{z} \mu_j\, dn_j.$$

Bearing in mind that the phase is inert, i.e. not chemically active, we may equally well understand dn_j to be the amount of C_j which appears within it. It is this latter interpretation which is to be retained in the presence of chemical reactions and without further ado we then take it for granted that our expression for dW continues to be valid.

The same generic consequences flow from the Zeroth, First and Second Laws whether the system be open or closed, in the sense that the existence of the energy U, the metrical entropy S and the absolute temperature T are guaranteed, though U and S are now of course functions of the internal coordinates as well. As before $TdS = dU + dW$, or explicitly,

$$TdS = dU + \sum_{k=1}^{n-1} P_k\, dx_k - \sum_{j=1}^{z} \mu_j\, dn_j.$$

Finally, granted that each phase is a standard system in its own right, a general system is simply a compound standard system of which the phases are the parts, and then

$$TdS = dU + \sum_{k=1}^{n-1} P_k\, dx_k - \sum_{a=1}^{\varphi} \sum_{j=1}^{z} \mu_{aj}\, dn_{aj},$$

$S\,(= \Sigma\, S_a)$ and $U\,(= \Sigma\, U_a)$ now referring to K as a whole. Except when K consists of a single chemically inert phase its internal coordinates are not mutually independent. Under conditions of equilibrium certain relations between them must be satisfied, just as the external coordinates of two standard systems in mutual diathermic contact had to satisfy an equilibrium condition. In short, the number of *degrees of freedom* f of K is less than $n + \varphi z$, but as a matter of convenience we defer the determination of the actual value of f to a later lecture.

We see incidentally that if K happens to be closed the expression for TdS cannot make any explicit reference to its internal constitution. This is only another way of saying that the internal state of a closed system is already determined by its external state. At any rate, consistency demands that $\Sigma\Sigma\mu_{aj}\, dn_{aj}$ must vanish; and we might anticipate that this condition should enable us to determine the internal state explicitly. In a sense this is so. By way of example, take K to consist of just one phase, governed by a single chemical reaction. Then there are integers ν_i, positive or negative, such that $dn_1/\nu_1 = dn_2/\nu_2 = \ldots = dn_z/\nu_z$ $(= d\xi$, say) [*Note* 25]. The vanishing of $\Sigma\mu_j\, dn_j$ therefore implies that $\Sigma\mu_j\,\nu_j = 0$. On the other hand, the system was prepared initially by assembling n_{10} moles of C_1, n_{20} moles of C_2, and so on, so that when final equilibrium obtains $n_j = n_{jo} + \nu_j\xi$. Granted that the $\mu_j\,(x_1, \ldots, x_n, n_1, \ldots, n_z)$ are known functions, the condition $\Sigma\mu_j\,\nu_j = 0$ therefore amounts to an algebraic equation for the value of the *degree of advancement* ξ. Given the external state, the internal state can therefore be found by calculation. The only snag is that, having determined one or several values of ξ in this way, we do not know whether these in fact correspond to physically realizable states. We therefore pursue this sort of argument no further and wait until we can tackle the problem by a different method.

The time is now opportune to turn our attention to certain auxiliary functions called *characteristic functions* (or also *thermodynamic potentials*). As the name suggests, any one of them, taken as a function of variables appropriate to it, completely characterizes the behaviour of the system. In the present context we shall suppose there to be only a single d-coordinate which we specifically take to be the volume V [*Note* 26]. It is sufficient for our purposes to single out the following prominent examples (i) the *Helmholtz potential* $F = U - TS$, (ii) the *Gibbs potential* $G = U - TS + PV$, (iii) the *grand potential* $X = PV$, all of which have the dimensions of energy. We consider them in turn.

Upon eliminating TdS from the total differential of F we find for a single phase that

$$dF = -SdT - PdV + \Sigma\, \mu_j\, dn_j.$$

Therefore, provided F is given as a function of T, V and the n_j [*Note* 27],

$$S = -\partial F/\partial T, \quad P = -\partial F/\partial V, \quad \mu_j = \partial F/\partial n_j,$$

and incidentally $U = F - T\partial F/\partial T$, a relation known as the *Gibbs-Helmholtz equation*. Thus F indeed generates the functions U, S, P by μ_j by mere differentiation, and between them these contain everything we can know about the thermodynamic behaviour of K.

Since the temperature of a system undergoing an irreversible transition is not defined we shall call such a transition of K isothermal if its surroundings are at a fixed temperature $\bar{T}$. We are reminded at once of the inequality $\triangle S \geqslant \int dQ/\bar{T}$ which we encountered earlier. Indeed, with $\bar{T}$ constant, this gives $\bar{T}\triangle S \geqslant \int dQ = \triangle U + W$, or $W \leqslant -\triangle F$. Because of this result, F is sometimes called the (Helmholtz) *free energy*. However, whereas in an adiabatic transition $W = -\triangle U$ always, in an isothermal transition the relation $W = -\triangle F$ is true only when reversibility obtains.

Next, since $G = F + PV$,

$$dG = -SdT + VdP + \Sigma\, \mu_j\, dn_j,$$

so that T, P and the n_j are the variables appropriate to G. Then

$$S = -\partial G/\partial T, \quad V = \partial G/\partial P, \quad \mu_j = \partial G/\partial n_j.$$

Now we can think of a system of volume V and mole-numbers n_j as being arbitrarily compounded out of two sub-systems K_A and K_B of volumes $(1 - \varepsilon)V$, εV and mole-numbers $(1 - \varepsilon)n_j$, εn_j respectively, P and T having the same values for both. Since U, S and V are additive, so is G, and therefore $G_A = (1 - \varepsilon)G$, $G_B = \varepsilon G$. But, on taking ε sufficiently small, $G_A - G$ is just given by dG with $dP = dT = 0$ and $dn_j = -\varepsilon n_j$. It follows that

$$G = \Sigma\, \mu_j\, n_j.$$

The differential of this, when compared with the earlier form of dG, gives the *Gibbs-Duhem identity*

$$SdT - VdP + \Sigma\, n_j\, d\mu_j = 0.$$

At the same time, G is evidently a homogeneous function of degree 1 of the n_j. In other words, the μ_j can only depend upon their ratios, or equivalently, on the concentrations $c_i = n_i/\bar{n}$, where $\bar{n} = \Sigma\, n_i$.

As regards X, we can now write $X = \Sigma\, \mu_j\, n_j - F$ and therefore

$$dX = SdT + PdV + \Sigma\, n_j\, d\mu_j$$

so that

$$S = \partial X/\partial T, \quad P = \partial X/\partial V, \quad n_j = \partial X/\partial \mu_j,$$

granted, of course, that X is given as a function of the variables T, V, $\mu_1, \ldots, \mu_z$. It is worth noting that since $X = PV$ by definition, the second equation shows at once that P does not depend upon V.

Various other characteristic functions are sometimes used, such as the *enthalpy* $H(S, P, n_j) = U + PV$, or $\tilde{F} = F/T$, and the like: the list can obviously be extended almost endlessly. However, the relevant equations and appropriate variables are evidently so easily found that there is not much sense in trying to memorize them. At any rate, given some particular characteristic function of each of the phases of a system, that of the system as a whole is just the sum of the individual characteristic functions, granted that each phase can be regarded as a standard system.

Problems

8.1 Can U be regarded as a characteristic function?

8.2 For the sake of simplicity, take the system as closed so that the internal coordinates can be ignored. Can $F(P, T)$ be regarded as a characteristic function?

8.3 In the context of the grand potential we arrived at the somewhat bewildering conclusion that P does not depend on V. How can that be?

8.4 Given the Helmholtz potential of a closed system, how do you obtain its equation of state?

8.5 Show that the Gibbs potential of n moles of an ideal gas whose equation of state is $PV = nRT$ has the generic form

$$G = n(RT \ln P + \chi(T))$$

and find an expression for $\chi(T)$.

LECTURE 9

The Sign of T. The Third Law

The time is now opportune to discuss the question of the sign of T. To begin with, consider two alternative infinitesimal isometric transitions between two given states of some system K. The first transition is to be adiabatic. It is therefore necessarily irreversible, so that $dS > 0$ whilst $dU > 0$ also, as we already know. The second, on the other hand, is to be reversible—meaning quasi-static—and so necessarily diathermic. In this case $dS = dU/T$, where dS and dU are the same quantities as those which we encountered a moment ago. It follows that $T > 0$. We must remember, however, that we are here excluding certain 'abnormal' systems for which some of our assumptions are not justified [cf. Lecture 20].

It is worth re-deriving this important result by a different route. In any reversible transition of K between states $\mathfrak{S}'$ and $\mathfrak{S}''$, say, $TdS = dU + Pdx$ for any element of it. x is some d-coordinate, and if there are any others we simply keep them fixed. Take x and the empirical temperature t as coordinates. Then the integrability condition [*Note* 28] on dS at once gives the relation $T^{-1}dT/dt = P_t/(P + U_x)$, with subscripts denoting partial derivatives. (This commonly used notation greatly improves the appearance of thermodynamic relations and we shall often use it later.) It follows that

$$T''/T' = \exp \int_{t'}^{t''} P_t \, dt/(P + U_x).$$

Inspection reveals that the absolute temperatures of these states, both of which were chosen arbitrarily, must have the same sign. Bearing in mind that $T(t)$ is a universal function, we now only need to find the sign of T for one state of one system to know it for all states of all systems. Accordingly we choose an ordinary gas such as oxygen under everyday conditions. Then experiment shows that when t is kept constant PV is (nearly enough) constant, and we may simply take $t = PV$. Further, U is a function of t alone, i.e. $U_V = 0$. It follows straight away that $T''/T' = t''/t'$, that is to say $T = bt$, where b is a constant. Then $dS = dU/T + dV/bV$, so that $S = \int dU(T)/T + b^{-1} \ln V$. Now, in any adiabatic expansion of the gas in which no work is done on it $Q = 0$ and $W = 0$ so that $\triangle U = 0$, i.e. $\triangle t = 0$. Hence $b\triangle S = \triangle \ln V > 0$. Since, on the other hand, the process is irreversible, $\triangle S > 0$. We conclude that b, and therefore T, must be positive. What has thus been clearly brought out is that the invariability of the sign of T is demonstrable by considering quasi-static processes alone, whereas the actual sign can be found only by drawing upon at least one specific irreversible process [*Note* 29].

The conclusion that T cannot be negative at once leads us to question the physical status of the temperature $T = 0$ which is absolutely distinguished from all others. On the face of it, we might well be confronted with a singularity of the integrand of $\int dQ/T$, yet we should not be too surprised if the behaviour of systems in fact turned out to be just such as to prevent the occurrence of this singularity. In that case the properties of functions such as the forces P_k would of course have to be very different from what we might have expected them to be on the basis of experimental results obtained under 'ordinary' conditions. Now, when thinking about systems at lower and lower temperatures we are at once faced by the difficulty of being unable to say in any precise sense when a given temperature is 'low': low compared with what? To say that 10^{-3} °K is 'extremely low' is presumably intended to convey that as the temperature of the system is reduced still further no new, unexpected phenomena come into play. Such an assumption is, however, dangerous. Thus, the *specific heat* C [*Note* 30] of a gas such as hydrogen which once appeared to be constant turned out to decrease with T below about 100 °K. As better refrigerators were developed the quite unexpected phenomenon of superconductivity was discovered, the transition temperatures of many elements being in the region 1 °K to 10 °K. Going well below this again, the specific heats of certain magnetic salts have pronounced maxima at temperatures of a few millidegrees. At any rate, these few examples amply illustrate the dangers of uncritically extrapolating to $T = 0$.

The question now is this: can we detect some common trend which runs through the results of experiments carried out at ever lower temperatures? To begin with, it turns out that as $T \to 0$ all specific heats go to zero, intermediate relative maxima notwithstanding, and moreover they do so at least linearly with T. This result is at any rate consistent with the assumption that U is a good function in a neighbourhood of $T = 0$. The convergence of the formal integral $S = \int_o^T dQ/T$ (evaluated for constant d-coordinates) is then assured [*Note* 31]. Next, consider reversible adiabatic transitions of a system in which only T and one d-coordinate x (such as the volume, or the component of a magnetic moment in the direction of an external magnetic field) are allowed to vary. Then $CdT + (U_x + P)dx = 0$, or, on setting $t = T$ in the equation for $T^{-1}dT/dt$ found earlier, $dT = -C^{-1}TP_T dx$. $C^{-1}T$ certainly does not tend to zero with T, as we have seen. It follows that a finite change of x would lead to the attainment of $T = 0$ unless P_T itself tends to zero with T. All attempts to cool systems to $T = 0$ in this way have in fact failed [*Note* 32], so that P_T must become zero in this limit. On the other hand, the integrability condition on dF is that $P_T = S_x$, and we thus conclude that

$$S_x \to 0 \text{ as } T \to 0.$$

It remains to enquire into the dependence of S upon the internal coordinates n_i, granted that K is closed. To clarify the situation it will suffice to consider the example of a system made up of only three constituents. Specifically K shall have been prepared by introducing one mole of iron, C_1, and one mole of sulphur, C_2. These will react to form iron sulphide, C_3. When equilibrium has been reached the amounts of the various constituents present will be $n_1 = n_2 = 1 - \xi$, $n_3 = \xi$, the value of the degree of advancement ξ being determined already by the external state. If, therefore, we wish to keep ξ under our control we have no option but to subdivide K physically into three compartments, mutually separated by means of fully flexible diathermic partitions, these compartments containing $1 - \xi$ moles of C_1, $1 - \xi$ moles of C_2 and ξ moles of C_3, respectively. Now, the transformation of C_1 and C_2 into C_3 can be carried out reversibly. For instance, C_3 is heated until it is sufficiently dissociated and then C_1 and C_2 are separated out by means of semi-permeable membranes, all constituents being finally cooled to the initial temperature of C_3. By measuring specific heats and heats of reaction we can therefore experimentally compare the entropies of states of K as a whole for different values of ξ. What one actually finds in this way is that as $T \to 0$ the entropy of K becomes independent of ξ.

Now, in this example ξ $(= n_3)$ operates as the only 'free' internal coordinate, and in the limit $T \to 0$ the entropy becomes independent of this and of all d-coordinates. That a corresponding state of affairs in fact universally obtains for general (standard) systems amounts to the *Third Law of Thermodynamics:*

the entropy function of a closed system has the same finite value for all states which have $T = 0$.

This statement has a formal character in that it is concerned with the behaviour of a function at $T = 0$; the system itself cannot be at the absolute zero. We can save ourselves the trouble of examining its more immediate implications here, for they were already more or less explicitly contained in the discussion leading up to its formulation. However, it is worth remarking that the Third Law does not implicitly require the variables whose values constitute a 'state' to be the standard set of coordinates; in particular one may adopt the forces in place of the d-coordinates, and then we conclude, for instance, that the isobaric coefficient of expansion $(= (\ln V)_T)$ must vanish as $T \to 0$.

Curiously enough, one occasionally encounters discussions of supposed 'exceptions to the validity of the Third Law'. There are, however, no exceptions; what is, in fact, being discussed in detail is the entirely unsurprising conclusion that the law is invalid under circumstances in which it does not claim to be valid. It, like the other three laws, is a statement about states and so presupposes equilibrium. However, it is just at sufficiently low temperatures, where the regime of the Third Law becomes paramount, that the attainment of equilibrium may take so long that in effect it will never take place at all, and the implicit assumption of equilibrium will then lead to discrepancies.

The following is a typical situation. Upon being cooled, certain substances may end up in alternative forms, depending upon uncontrollable features of the experimental procedure. Specifically, a substance may then be in either an amorphous form or else possess a definite, regular crystalline structure. 'True' equilibrium corresponds to the latter, yet when the temperature is sufficiently low, the amorphous condition may, for all practical purposes, persist indefinitely. Everything appears straightforward: we could simply refuse to consider the behaviour of the amorphous substance in the context of the Third Law. In doing this we would, however, throw out the baby with the bathwater, for experiment shows that the Third Law can safely be taken to remain relevant, provided that at the end of the experiment the substance is still (or again) amorphous. It thus behaves, indeed, as if it were in equilibrium. There is, however, nothing to say that the value S_0 which S has for all states at $T = 0$ is the same as that which it would have had if the substance had ended up as a crystal. Contradictions arise merely as the result of arbitrarily ignoring the difference $\triangle S_0$ between these 'zero point entropies', a difference which is certainly measurable.

As a matter of fact, we are evidently concerned here with a situation in which one of the defining conditions of standard systems is not satisfied in as far as a substance is present whose properties does depend upon its past history, albeit in a way which merely results in a dichotomy. Small wonder, then, that special care has to be exercised in the interpretation of experimental results.

Problems

9.1 Show that there are no substances whose equation of state is exactly $PV = RT$ (R = constant).

9.2 The energy density U/V of a certain 'fluid' is three times its pressure P. The fluid has the further unusual property that $P(V, T)$ is in fact independent of the volume. Is the Third Law violated?

9.3 Under a wide range of conditions experiment shows that for the rare gases $U = \frac{3}{2}PV$. Show that any gas which is strictly subject to this relation must have an equation of state of the form $PV = Tg(z)$, where $z = TV^{\frac{2}{3}}$ and g is an undetermined function. (Note that this result can accommodate the ideal gas equation $PV = RT$, namely, with $g(z)$ = constant.) Show also that the Third Law will not be violated if the specific heat C goes to 0 with T as some positive power n of T (the condition $n \geqslant 1$ is therefore not necessary here!), but that the mere vanishing of C is not a sufficient condition for the Third Law not to be violated.

LECTURE 10

Pseudo-states. Equilibrium Conditions. The Phase Rule

By and large we have now reached the stage at which we have disposed of the basic issues of the phenomenological theory. Nevertheless some matters of general principle remain to be discussed. In particular we have yet to answer two related questions raised earlier: how is the internal state of a closed physico-chemical system K to be found, given its external state, and what is the number of degrees of freedom of such a system, now regarded as open? In this context it is helpful to introduce the notion of *pseudo-states* corresponding to a given state of K.

Consider, then, together with a given isolated standard system K, another isolated system K^* which is a modified version of K in the sense that it is like K except for the presence of additional internal partitions of one kind or another and possibly of 'anti-catalysts', i.e. of substances whose sole purpose is to prevent the occurrence of otherwise possible chemical reactions. Now let K^* be in a state $\mathfrak{S}^*$. Upon removing the additional partitions and anti-catalysts we effectively end up with the system K and this will be in a certain state $\mathfrak{S}$. Then $\mathfrak{S}^*$ was a pseudo-state of K corresponding to the state $\mathfrak{S}$ of K.

By way of example, we return once again to our sample system, taking V and T as its coordinates. K^* might then be a system which differs from K by containing an additional fixed internal adiabatic partition. The volumes V_A, V_B of the two compartments K_A, K_B of which K^* consists being fixed, only temperatures need here be considered. Then $\mathfrak{S}^*[T'_A, T'_B]$ is a pseudo-state of K corresponding to the state $\mathfrak{S}\,[T']$ of K if the state attained upon removal of the partition has the temperature T'. This will be so if only the condition $U_A(V'_A, T'_A) + U_B(V'_B, T'_B) = U(V', T')$ is satisfied.

Now, according to the Entropy Principle, if S is the entropy of a state $\mathfrak{S}$ of K and S^* is the entropy of a corresponding pseudo-state $\mathfrak{S}^*$ then $S^* \leqslant S$. This immediately suggests the following general principle for the characterization of equilibria: *the entropy of any state of an isolated system is not less than the entropy of any corresponding neighbouring pseudo-state.* (Here 'neighbouring' means that $|S^* - S|$ shall be sufficiently small.)

In view of the isolation of K the energy U^* of any pseudo-state $\mathfrak{S}^*$ must have the same value as the energy U of the state $\mathfrak{S}$. In short, the *equilibrium conditions* can be written

$$dS^* = 0, \qquad d^2S^* < 0,$$

granted that U^* and the d-coordinates are kept fixed [*Note* 33]. Of course, all constraints inherent in the closedness of K and in the presence of chemical reactions, if any, must also be taken into account. Occasionally the term 'equilibrium condition' is taken to refer merely to the requirement for an extremum of S^*. The inequality $d^2S^* < 0$, which ensures that the extremum is in fact a maximum, then goes under the name of *stability condition*.

(Should it happen that $d^2S^* = 0$ when $dS^* = 0$, then the dominant term of $S^* - S$ must be negative for pseudo-states $\mathfrak{S}^*$ sufficiently close to $\mathfrak{S}$.)

Hitherto we have prescribed that U^* 'be constant' ($= U$), that is to say, that all pseudo-states have the same energy as that of the state to which they correspond. Suppose that we consider instead the constancy of T^* ($= T$), by which we mean precisely that K, instead of being adiabatically isolated, is now in diathermic contact with its surroundings at the fixed temperature T. Modification of K followed by restoration of its initial structure allows us again to contemplate transitions of K^* from pseudo-states corresponding to a given state $\mathfrak{S}$ of K. As we already learned in the eighth lecture, the change in the Helmholtz potential F^* is less than the work done on K by its surroundings. Here, however, no work is done since the d-coordinates are required to be kept fixed. Hence F^* decreases and the equilibrium conditions are

$$dF^* = 0, \qquad d^2F^* > 0.$$

The variations are of course subject to the conditions we encountered in the context of S^* except that the constancy of T^* takes the place of the constancy of U^*. Since we are only contemplating small variations, the new conditions are entirely equivalent to the old.

What we have done in effect is to go over from the characteristic function S to the characteristic function F; and we are not prevented from going over to any other characteristic function of our choice. In particular G commends itself in practice, K now being required to be in diathermic contact with its surroundings (as in the case of F) whilst the forces P_i are kept fixed. Then we have, equivalently to the earlier conditions,

$$dG^* = 0, \qquad d^2G^* > 0.$$

We are at last in a position to determine the number of degrees of freedom of a given system K. Even though it is open we must take it as closed whilst examining the consequences of the equilibrium conditions since K will only be in a specific state whilst it is left to itself. For the time being we take it to be inert. Let the pseudo-states corresponding to a state $\mathfrak{S}$ of K differ from it only by virtue of the $n_{\alpha j}$ not having the values implicit in $\mathfrak{S}$. Then the condition $dG^* = 0$ at once requires $\Sigma\Sigma\mu_{\alpha j}dn_{\alpha j} = 0$. However, the total amount of any constituent present in K is fixed, so that the $dn_{\alpha j}$ are subject to the z restrictions $\sum_{\alpha} dn_{\alpha j} = 0$. If the jth of these be supplied with a Lagrange multiplier λ_j the equilibrium condition becomes $\Sigma\Sigma(\mu_{\alpha j} - \lambda_j)dn_{\alpha j} = 0$, and the $dn_{\alpha j}$ are now to be treated as if they were independent from each other. It follows that $\mu_{\alpha j} = \lambda_j$, which means that the chemical potentials of any given constituent have the same values in all the phases.

Now, states of an open system are prepared by suitably adjusting the n external coordinates and injecting appropriate amounts of the various constituents. Any such state must satisfy the $z(\varphi - 1)$ equations $\mu_{\alpha j} = \lambda_j$. On the other hand, altogether $n + z\varphi$ variables x, $n_{\alpha j}$ enter into the description of K. We conclude that in aiming at the preparation of such and such a state we are in fact only free to specify the values of $n + z$ of these $n + z\varphi$ variables at will, and in this sense K has $n + z$ degrees of freedom. However, we are normally not interested in the total masses of the various phases [*Note* 34]. We therefore agree to the convention that in counting the number of degrees of freedom f the total masses of the φ phases are now to be omitted. Accordingly f is given by the relation

$$f = z - \varphi + n,$$

a result which is usually referred to as the *(Gibbs) phase rule*. Its content, put slightly differently, is simply this: arbitrary values can be assigned to only $z - \varphi + n$ of the $n + (z - 1)\varphi$ intensive variables P_k, T, μ_{aj}.

We yet have to allow for the possibility of chemical reactions being present. We recall that the equation of any such reaction in the αth phase is $\Sigma \nu_{aj} C_j = 0$. Therefore, if there are R independent reactions altogether, we have R equations $\Sigma_j \nu_{aj}^{(r)} C_j = 0$ $(r = 1, \ldots, R)$. (Independence means that one cannot find R numbers γ_r such that $\Sigma_r \gamma_r \nu_{aj}^{(r)} = 0$ simultaneously for all j.) As a consequence of the rth reaction taken by itself $dn_{aj}^{(r)} = \nu_{aj}\, d\xi^{(r)}$, where $\xi^{(r)}$ is the rth degree of advancement. Altogether any dn_{aj} is made up of r contributions of this kind, together with a contribution $\bar{d}n_{aj}$ representing the mere transfer of C_j between the phases. The $\bar{d}n_{aj}$ are again subject to the restrictions $\Sigma \bar{d}n_{aj} = 0$ which are accommodated by means of Lagrange multipliers λ_j as before. The equilibrium condition $dG^* = 0$ therefore now reads

$$\sum_a \sum_j [\mu_{aj} \sum_r \nu_j^{(r)} d\xi^{(r)} + (\mu_{aj} - \lambda_j)\bar{d}n_{aj}] = 0$$

in which the $\xi^{(r)}$ and $\bar{d}n_{aj}$ are to be treated as independent of each other. It follows that in addition to the previous $z(\varphi - 1)$ equations $\mu_{aj} - \lambda_j = 0$ the R equations $\Sigma\Sigma \mu_{aj} \nu_{aj}^{(r)}$ must also be satisfied. Therefore now

$$f = z - \varphi + n - R,$$

and this is the phase rule in the presence of chemical reactions. Occasionally further conditions must be satisfied by the concentrations, for example in the case of electrolytic dissociation. One can take the view that if there are R' such conditions and R chemical reactions then only $\bar{z} = z - R - R'$ of the z constituents are independent of each other. Then

$$f = \bar{z} - \varphi + n.$$

When a system consists merely of one inert constituent C_1 (external coordinates P and T, say) we have $z = 1$, $n = 2$ and so $f = 3 - \varphi$. Consequently more than three phases can never coexist, and three phases can coexist only in a single definite state, usually called the *triple point* of C_1. It is that pair of values of P and T which satisfies the pair of simultaneous equations $\mu_{11} = \mu_{21} = \mu_{31}$. This, then, is a very rudimentary illustration of the operation of the phase rule. Bearing in mind that whole treatises have been written about the implications of the phase rule, any discussion of more complicated situations would clearly be out of place. On the other hand, we must not forget that any determination of actual equilibria must include a verification that the stability condition is in fact satisfied. In other words, we have to demonstrate that $\Sigma(\partial^2 G^*/\partial n_{ai} \partial n_{\beta j}) dn_{ai} dn_{\beta j} > 0$ for all variations $dn_{\gamma k}$ consistent with the constraints imposed upon them by the closedness of the system and by the occurrence of chemical reactions. These variations must include those which correspond to the appearance of phases absent from a state which satisfies the condition $dG^* = 0$.

Problems

10.1 A system consists of a single inert constituent in an enclosure which contains no internal partitions. Can there in principle be a state in which four phases coexist?

10.2 During the lecture a set of pseudo-states of our sample system was described. Show that the stability condition is always satisfied.

10.3 A movable rigid diathermic partition divides an enclosure of volume V, containing a single inert fluid, into compartments of volume V_A, V_B. By considering pseudo-states at fixed V, T, corresponding to various positions of the partition, examine the consequences of the appropriate conditions $dF^* = 0$, $d^2F^* > 0$, granted that there is only a single phase.

10.4 K is a one-phase system of z constituents between which there is one chemical reaction. Having first verified that $\Sigma n_i g_{ij} = 0$ always ($g_{ij} = \partial^2 G/\partial n_i \partial n_j$) show that the stability condition is certainly satisfied if $g_{ij} < 0$ ($i \neq j$), but that this is not a necessary condition for stability.

10.5 Show that in the course of boiling a solution of rock salt in water at constant pressure P whilst crystallization of the salt is taking place, its concentration c and the temperature T remain constant.

LECTURE 11

The Idea of 'Applications of Thermodynamics'

Since the main purpose of this course of lectures is to elucidate the underlying notions of thermodynamics, allowing physical insight to take precedence over mathematical niceties, it would be out of place here to present a large variety of possible applications of the theory. These may be found elsewhere [*Note* 35]. Instead, we shall do well first to clarify just what we mean by 'applications'; and, having done so, to illustrate the points at issue by considering a few characteristic examples which will also show how simple it all is in principle.

To begin with, discussions of the phase rule for instance are often undertaken in the context of 'applications', whereas we plainly regarded it as a basic part of the theory. Now, it is of course permissible to develop a theory to a certain stage at which one simply declares its basic part to have been disposed of, and then to call all further consequences of the theory applications of it, no matter what further assumptions may have had to be introduced. This, however, implies an arbitrariness which easily leads to conceptual confusion. We therefore prefer to take the view that the theory as such contains all those results which can be derived without drawing upon any information about the properties of specific systems; information implicit in the actual form which functions, such as the thermodynamic potentials, may happen to have in this or that particular case. If we nevertheless insist on calling some results of this kind 'applications'—for the sake of tradition—we shall wisely refer to them as 'formal applications'; whereas 'proper applications' will be concerned with specific systems, or specific classes of systems.

Formal applications, then, rest (i) upon the mere fact of the existence of certain functions (guaranteed by the Zeroth, First and Second Laws), together with the fundamental differential relation $TdS = dU + dW$ which connects them with each other; (ii) upon generic restrictions which the Third Law imposes upon their behaviour; and (iii) upon the Entropy Principle, particularly as embodied in the equilibrium conditions. For the purposes of illustration it suffices to consider systems with only two external coordinates, one of which is its volume V, the corresponding force being the pressure P. Then, on the most rudimentary level, the existence of an empirical temperature function $t = t(V, P)$ alone implies the relation $\beta = a/P\kappa$ between the isometric pressure coefficient β, the isobaric coefficient of expansion a and the isothermal compressibility κ [*Note* 36]. Again, Joule expansion, i.e. the adiabatic expansion of a fluid without the performance of work is characterized by the constancy of U, so that we have the algebraic relation $U(V + \triangle V, T + \triangle T) - U(V, T) = 0$ between $\triangle V$ and $\triangle T$.

Next we have the class of differential relations which explicitly or implicitly merely express the fact that $(dU + dW)/T$ is a total differential. To preserve generality, let the coordinates (not necessarily all external) indiscriminately be denoted by $y_1, y_2, \ldots$; and let J be some function of these. dJ, rewritten in any way by means of the relations $dU + dW - TdS = 0$, takes the form $\Sigma Y_k dy_k$, say. Then the integrability conditions $\partial Y_i/\partial y_j = \partial Y_j/\partial y_i$ $(i, j = 1, 2, \ldots)$ [recall *Note* 28] are just the differential relations in question. For example taking $J = U, F, G, H$ in turn, with $(y_1, y_2) = (S, V), (T, V), (T, P), (S, P)$ respectively, we straight away get *Maxwell's relations*

$$P_S = -T_V, \qquad P_T = S_V, \qquad V_T = -S_P, \qquad V_S = T_P.$$

Similarly, when $J = F/T$ and $(y_1, y_2) = (V, T)$, we find that $U_V/T^2 = (P/T)_T$, a result which we already encountered earlier. We may use it to deduce in turn that Joule expansion is governed by the equation $\triangle T/\triangle V = (P - TP_T)/C$, when $\triangle T$ and $\triangle V$ are sufficiently small.

Another kind of formal application which falls into the present context revolves about the discontinuities of characteristic functions implicit in the existence of phase-changes. Suppose, indeed, that a certain substance at pressure P undergoes a (reversible) change from the first to the second phase at the temperature $T_\varphi(P)$. Denoting molar quantities by lower case symbols, the quantity $\triangle g \equiv g_1(P, T_\varphi) - g_2(P, T_\varphi)$ vanishes in equilibrium, as we already know. Differentiation gives $\partial \triangle g/\partial P = 0 = \triangle \mathrm{v} - (\triangle s)dT_\varphi/dP$, or, if $\lambda = T\triangle s$ is the *latent heat* of the phase change,

$$dP/dT_\varphi = \lambda/(T_\varphi \triangle \mathrm{v}),$$

which is the *Clapeyron-Clausius equation.* Evidently we have here contemplated the situation in which $g(P, T)$ is continuous whilst its first derivatives are not ($\triangle s \neq 0, \triangle \mathrm{v} \neq 0$), that is, we have what is called a *first order transition.* In an sth order transition g and its derivatives to order $s - 1$ are continuous but those of order s are not.

In the discussion preceding the formulation of the Third Law we already encountered formal applications in which this law was involved, and there is no need to repeat them. We merely remind ourselves that the differential relations which we discussed a short while ago—typically the second and third Maxwell relations—at once lead to statements about the behaviour, as $T \to 0$, of functions other than S. Further, the independence of S of degrees of reaction as $T \to 0$ is a necessary ingredient of the theoretical determination of actual chemical equilibrium. Finally, we have those formal applications which draw also upon the equilibrium conditions. They, however, are already adequately illustrated by our earlier derivation of the phase rule and of various concomitant results.

It remains to say a few words about proper applications. All of these represent inferences about the behaviour of particular systems whose structure and constitution is reflected in the specific forms of those functions which enter only generically into the formal applications. The range of situations to be contemplated is limitless. As a matter of practical realities no less than as a matter of principle we therefore must rest content with a few simple examples, all of which merely relate to gases, and for the most part we shall even take these to be ideal or nearly ideal.

To begin with, a system consisting of n moles of one ideal gas C_1 has the equation of state $PV = nRT$, from which it follows that its Gibbs function has the form $G = n[\chi(T) + RT \ln P]$. The function χ is determined by the equation of state and the measured specific heat only to within an additive linear function of T. By differentiation, if a dot denotes differentiation with respect to T

$$V = nRT/P, \qquad S = -n(\dot{\chi} + R \ln P), \qquad U = n(\dot{\chi} - T\chi - RT).$$

This is as must be since the correct equation of state is reproduced whilst U depends on T only, so that the relation $U_V = T^2(P/T)_T$ is not violated. Of course, although under 'normal' conditions real gases are nearly ideal (with the same value of R for all of them) no substance can be ideal in the strict sense of the term since P_T does not vanish when $T = 0$, a result in conflict with the Third Law.

Next, the equation of state of a mixture of z ideal gases is $PV = \bar{n}RT$ and its energy is the sum of the energies of the separate constituents. We may therefore take it for granted that each constituent gas behaves as if all others were absent, C_j thus being in the mixture at temperature T and pressure $P_j = c_jP$. The Gibbs function of the system is therefore $G = \Sigma\, n_j(\chi_j + RT \ln P_j)$ and this shows at once that the chemical potential of C_j in the mixture is

$$\mu_j = g_j + RT \ln c_j,$$

where $g_j(P,T)$ is its molar Gibbs function [*Note* 37]. More generally a phase is called ideal if $\mu_j = g_j^*(P,T) + RT \ln c_j$ for every j, where g_j^* need no longer depend solely on the properties of C_j. A heterogenous *ideal system* then consists of ideal phases alone.

When a gas is non-ideal its equation of state may be exhibited as a power series, the *virial expansion* for example:

$$PV/RT = 1 + \sum_{r=1}^{\infty} B_{r+1}(T)V^{-r},$$

V being here the molar volume. Occasionally one also considers equations of state in closed form, such as the *van der Waals equation*

$$(P + aV^{-2})(V - b) = RT,$$

where a and b are positive constants, though such equations are useful chiefly for their heuristic value [*Note* 38].

Every result which now follows by inserting one equation of state or another into the generic relations obtained previously constitutes a proper application. For example, retaining only the first two terms of the virial series, Joule expansion leads to a change of temperature given by $\triangle T = -[RT^2V^{-2}C^{-1}dB_2/dT]\triangle V$, granted that $\triangle V$ is sufficiently small. An illustration a good deal richer in content is concerned with the implications of the equilibrium conditions upon a chemically active ideal system. It will suffice to consider a single phase in the presence of a single chemical reaction. In that case, the vanishing of $\Sigma\mu_i\nu_i$ immediately shows that the equilibrium concentrations are subject to the condition *(Law of Mass Action)*

$$\prod_{i=1}^{z} c_i^{\nu_i} = K_c(P, T), \qquad (K_c = \exp(-\Sigma\, \nu_i\, g_i^*/RT)),$$

where the *equilibrium constant* K_c of course depends on P and T only. In the particular case of a mixture of ideal gases, so that $g_i^* = g_i = \chi_i(T) + RT \ln P$, we have $K_c = P^{-\nu} \times \exp(-\Sigma\, \nu_i\chi_i/RT)$, with $\nu = \Sigma\, \nu_i$. In a region of temperatures in which the gases happen to have constant specific heats we have the remarkably detailed result $K_c = kP^{-\nu}\, T^a e^{\beta/T}$, where k, a, β are certain constants; for example $a = \Sigma\, \nu_i c_i^*/R$.

One would naturally like to be able to determine K_c directly, that is to say, by means of calorimetric measurements alone. How is this to be done? To begin with, the s_i can be obtained by calculation to within an unknown additive constant s_{oi} once the specific heat at constant pressure c_i^* has been measured [*Note* 39]: $s_i = s_{oi} + \int_o^T (c_i^*/T)dT$. Next, by definition, $R \ln K_c = -\Sigma\, v_i g_i/T = \Sigma\, v_i\, [s_i - (u_i + Pv_i)/T]$. On the other hand, in the course of a reaction in which $v_1, v_2, \ldots, v_z$ moles of the various gases appear, an amount of heat Λ—the *heat of reaction*—is produced which, since the whole process takes place at constant pressure, is just given by $-\Sigma\, v_i\,(u_i + Pv_i)$. Thus, $R \ln K_c = \Lambda/T + \Sigma v_i \times \int(c_i^*/T)dT + \Sigma v_i\, s_{oi}$. Here only the value of $\Sigma v_i\, s_{oi}$ appears to be unknown. Happily, as we recall from the discussion which led up to the statement of the Third Law, the latter requires this value to be zero, so that K_c can indeed be measured calorimetrically.

By measuring equilibrium concentrations one can check the validity of the Law of Mass Action: it frequently turns out that one should in fact attach a non-zero value to $\Sigma v_i\, s_{oi}$. This was to be expected; for in measurements of specific heats at low temperatures substances will frequently not be in equilibrium, even though they appear to be so. This point we discussed, however, already at the end of the ninth lecture, and we hardly need go over the same ground again, except to say that we now have a practical method for measuring the differences between the values of s_{oi} which correspond to different modifications in which C_i may end up as $T \to 0$.

Problems

11.1 Show that if C and C^* are the specific heats at constant volume and constant pressure respectively,

$$C^* - C = T\, P_T\, V_T\, (= VT^2 a^2/\kappa).$$

11.2 Derive the third Maxwell relation, and show also that $\partial\mu_j/\partial T = -\partial S/\partial n_j$.

11.3 Derive *Clausius' equation* for the difference between the molar specific heats c^* of a substance in phases 1 and 2.

11.4 Show that the saturated-vapour pressure varies with temperature according to $P = \text{constant} \times \exp(-\lambda/RT_\varphi)$, granted that λ is constant and the molar volume of the liquid is negligible compared with that of the vapour.

11.5 One mole of hydrogen combines with one mole of iodine to form two moles of hydrogen iodide. How does the equilibrium constant K_c for this reaction vary with the pressure?

LECTURE 12

Pseudo-states Revisited: Phenomenological Theory of Fluctuations. Ensembles

The introduction of the notion of pseudo-states was originally motivated by the wish to give concrete expression to the view that formally the stability of systems must be connected with the Entropy Principle. After all, if an isolated standard system K were not in a stable condition it would eventually attain a certain state $\mathfrak{S}'$ whose entropy S' would be greater than that of the initial 'state' $\mathfrak{S}$. The reason for using quotation marks just now is this: K was in fact initially not in a state at all and no entropy could be ascribed to it. The interpretation of $\mathfrak{S}$ as a pseudo-state, however, makes everything well-defined.

So far so good; yet we might well begin to worry about the following question. Suppose K is in a given state $\mathfrak{S}$, then—somewhat naïvely expressed—how does K know that the entropy S^* of every neighbouring pseudo-state $\mathfrak{S}^*$ corresponding to $\mathfrak{S}$ is less than that of $\mathfrak{S}$? It was all very well for us to make pseudo-states physically realizable by suitably modifying K, but K, left to itself, 'isn't aware' of what might be the case under other circumstances. Happily nature comes to our rescue at this point. Close inspection reveals that, contrary to what has hitherto been taken for granted, in the absence of any interference on our part no system is quiescent, even though we are still operating on a phenomenological scale. For example, the density of a gaseous system is not strictly uniform but, in a neighbourhood of any selected point, fluctuates about a mean value. The energy of a sub-system of K likewise fluctuates about a mean value; and so on. In short, the system, nominally in a state $\mathfrak{S}$, of itself effectively explores neighbouring conditions, any such neighbouring condition being equivalent to a pseudo-state corresponding to $\mathfrak{S}$ [*Note* 40].

Having satisfactorily disposed of one problem we are now faced with another question: does the essential lack of quiescence of thermodynamic systems not make nonsense of everything we have done hitherto, right back to the very definition of 'state'? The answer is yes, but only if we obstinately refuse to admit a significant re-interpretation of the meaning of 'the value of' a quantity $\hat{A}$ which, under given circumstances, is capable of fluctuating. Ignoring the problem of experimental errors, we have hitherto taken the value of $\hat{A}$ to be the result A of a single measurement of $\hat{A}$. Henceforth, however, we shall instead understand it to be the mean $\bar{A}$ of the results A obtained in a large number of repeated measurements of $\hat{A}$. (All of these are of course to be carried out according to the same generic prescription, for it is just this prescription itself which defines the quantity $\hat{A}$ in the first place.) At this point it is worth remarking that on the whole it is wise to consider only fluctuations of extensive quantities since the very interpretation of the fluctuations of intensive quantities is already apt to lead to difficulties.

Now the mean $\bar{A}$ which we have just talked about is a temporal mean, in as far as it is obtained from a large number, ν, of measurements on a given system K, carried out sequentially. Conceptually it is, however, often simpler to contemplate instead another kind of mean, called the *ensemble mean*, as follows. We imagine a very large number, ν, of copies of K to have been constructed, each such copy being prepared consistently with whatever information is at hand about K itself. If, for instance, K has a given volume and physico-chemical composition, and is in diathermic equilibrium with its surroundings at temperature T then the same is true of every copy. K together with its copies is called a *representative ensemble* $\mathscr{K}$ of K appropriate to the given information. The value of $\hat{A}$ is now understood to be the mean $\bar{A}$ of all results A obtained by measuring $\hat{A}$ simultaneously on all members of the ensemble $\mathscr{K}$. (We take it for granted that the temporal mean and the ensemble mean are numerically equal, at any rate as $\nu \to \infty$.) Concomitantly we understand by a 'single measurement' of $\hat{A}$ a measurement of $\hat{A}$ carried out on a randomly selected member of $\mathscr{K}$.

The following question naturally springs to mind: what is the probability $\psi(A)dA$ that the result of a single measurement of $\hat{A}$ will lie in the range $(A, A + dA)$? For the moment we suppose $\hat{A}$ to be the only quantity capable of fluctuating. K itself is to be a standard system and so is the composite system K_+ made up of K and its surroundings $K^\dagger$, consistently with the idea that the fluctuations take place in the presence of thermal interaction between K and $K^\dagger$ [*Note* 41]. Now, having obtained some value A as the result of a single measurement of $\hat{A}$, we simply say that at the time of measurement the member of $\mathscr{K}$ in question, taken together with its surroundings, was exploring a pseudo-state $\mathfrak{S}_{\ddagger}^{*}$ in which $\hat{A}$ had just the value A. Unless A happened to be $\bar{A}$ the entropy $S_{\ddagger}^{*}$ of $\mathfrak{S}_{\ddagger}^{*}$ is less than the entropy S_+ of that state $\mathfrak{S}_+$ which satisfies the equilibrium conditions. Nothing is essentially altered if we admit the possibility of several quantities $\hat{A}$, $\hat{B}$, . . . being able to fluctuate simultaneously, in which case $\psi(A, B, \ldots)dAdB \ldots$ is the probability that the result of a single simultaneous measurement of $\hat{A}$, $\hat{B}$, . . . gives values in the ranges $(A, A + dA)$, $(B, B + dB)$, ... Here again, because of the equilibrium conditions, the absence of all fluctuations corresponds to the vanishing of the quantity $\sigma = S_+ - S_{\ddagger}^{*}$, and we suspect a close relationship between $\psi(A, B, \ldots)$ and σ. We therefore make the basic assumption that ψ depends upon $A, B, \ldots$ only through σ, that is to say, that there exists a function p of σ such that $\psi = p(\sigma)$.

What can we say about the function p? Certainly, since it must reflect the increasing rarity of larger and larger fluctuations, its value must fall off rapidly with increasing values of σ. However, we can do much better than this. We consider a composite system K made up of sub-systems K_1 and K_2 in mutual diathermic contact [*Note* 42] and suppose this to be of a kind such that the effect of the surroundings completely predominates over the effect of each sub-system on the other, as far as fluctuations are concerned. This implies that the fluctuations in K_1 and K_2 are statistically independent, and so their probabilities are multiplicative: $\psi(A_1, B_1, \ldots, A_2, B_2, \ldots) = \psi_1(A_1, B_1, \ldots)\psi_2(A_2, B_2, \ldots)$. Our previous assumption, taken together with the additivity of entropy, then shows at once that the functional relationship $p_1(\sigma_1)p_2(\sigma_2) = p(\sigma_1 + \sigma_2)$ must be satisfied. Therefore [*Note* 43]

$$p_j = \kappa_j e^{-\sigma_j/k} (j = 1, 2),$$

where κ_j is a constant such that $\int p_j \, dA_j \, dB_j \ldots = 1$, and k is a universal positive constant.

We can now return to the case of a single system K, and we suppose all fluctuations to be relatively very small. Writing $A_1, A_2, \ldots$ in place of $A, B, \ldots$ and $a_i = A_i - \bar{A}_i$, we need only retain the dominant terms of σ. On account of the equilibrium conditions, σ contains no terms linear in the a_l and hence we have generically [*Note* 44]

$$\psi(A_1, A_2, \ldots) = \kappa \exp(-\sum_{i,j} \alpha_{ij} a_i a_j),$$

where the α_{ij} do not depend upon the a_l. A convenient overall measure of fluctuations of individual quantities and of correlations between them is provided by the *bilinear means* (or *second moments*) $\triangle_{ij} = \overline{a_i a_j} = \kappa \int a_i a_j \exp(-\sigma/k) da_1 da_2 \ldots$. In particular $\sqrt{\triangle_{ii}}$ ($= \delta A_i$, say) is the root mean square fluctuation of A_i. The $\triangle_{ij}$ may be obtained merely by differentiation once the integral $J = \int \exp(-\sigma/k) da_1 da_2 \ldots$ has been evaluated. Thus, since $\int \psi \, da_1 da_2 \ldots = 1$ in the nature of things, we have $\kappa = 1/J$ and, by inspection,

$$\triangle_{ii} = -\partial \ln J/\partial \alpha_{ii}, \qquad \triangle_{ik} = -\tfrac{1}{2} \partial \ln J/\partial \alpha_{ik} \; (i \neq k).$$

If r is the number of fluctuating variables, $J = (\pi^r/j)^{\frac{1}{2}}$, where j is the determinant of the α_{ik} [*Note* 45].

The simplest example is provided by the energy fluctuations of our sample system when its volume is kept fixed. Then $\sigma = S(U, V) + S^\dagger(U^\dagger, V^\dagger) - S(U + u, V) - S^\dagger(U^\dagger - u, V^\dagger)$, bearing in mind that $U^\dagger + U$ has a fixed value. To the required order, $\sigma = \frac{1}{2}(\partial^2 S/\partial U^2 + \partial^2 S^\dagger/\partial U^{\dagger 2})u^2 = \frac{1}{2}(C^{-1} + C^{\dagger -1})T^{-2}u^2$ [cf. Problem **10.2**], the linear terms cancelling out, as they must, because $T = T^\dagger$. $K^\dagger$ is naturally taken to be very large compared with K, so that $C^{\dagger -1}$ is negligible compared with C^{-1}. Thus $j = \alpha_{11} = (2kCT^2)^{-1}$, so that $\triangle_{11} = \frac{1}{2}\partial \ln j/\partial \alpha_{11} = 1/2j = kCT^2$, or

$$\delta U = (kC)^{\frac{1}{2}} T.$$

When the volume can fluctuate as well, $j = \alpha_{11}\alpha_{22} - \alpha_{12}^2$, so that $\triangle_{11} = \alpha_{22}/2j$, $\triangle_{22} = \alpha_{11}/2j$, $\triangle_{12} = -\alpha_{12}/2j$. Before we go on to find explicit expressions for the α_{ij} we once and for all make allowance for the fact that $K^\dagger$ is much larger than K. First, let $\hat{X}$ be some quantity and X'', X' its values formally calculated for a given pseudo-state on the one hand and the state to which it corresponds on the other. Then write $\triangle X = X'' - X'$, so that $\sigma = -\triangle S_+ = -\triangle S - \triangle S^\dagger$. Since $K^\dagger$ is much larger than K the values of its intensive variables will not be sensibly affected by a fluctuation in K. Here therefore $-\triangle S^\dagger = -\int (dU^\dagger + P^\dagger dV^\dagger)/T^\dagger = -(\triangle U^\dagger + P^\dagger \triangle V^\dagger)/T^\dagger = (\triangle U + P\triangle V)/T$, so that

$$\sigma = -\triangle S + (\triangle U + P\triangle V)/T \; (= \triangle G/T),$$

an expression which makes reference to K alone.

Previously we used U as the one independent variable. We might equally well have used T: recall that any actual measurement takes place after the selected member of $\mathscr{K}$ has been isolated. When V is not fixed, it is, however, formally simpler to contemplate in the first place simultaneous fluctuations of T and V rather than of U and V. Then, with $A_1 = T$, $A_2 = V$, $2k\alpha_{ij} = -\partial^2 S/\partial A_i \partial A_j + T^{-1}\partial^2 U/\partial A_i \partial A_j$. By differentiation of the relation $TS_T - U_T = 0$ we find at once that $\alpha_{11} = C/2kT^2$, $\alpha_{12} = 0$. Likewise $TS_V - U_V = P$, so that $\alpha_{22} = -P_V/2kT$. Therefore finally

$$\delta T = \sqrt{\triangle'_{11}} = (k/C)^{\frac{1}{2}} T, \qquad \delta V = \sqrt{\triangle'_{22}} = (-kTV_P)^{\frac{1}{2}}, \qquad \triangle'_{12} = 0,$$

the primes serving as a reminder that we are dealing with fluctuations of T, not U. However, from $\triangle U = C\triangle T + U_V\triangle V + \ldots$ we infer at once that

$$\triangle_{11} = C^2\triangle'_{11} + U_V^2\triangle'_{22}, \qquad \triangle_{12} = U_V\triangle'_{22}.$$

Therefore, when V can fluctuate $(\delta U)^2$ has the additional term $-kTU_V^2\, V_P$.

Problems

12.1 The equation of state of a certain system is $PV^{\frac{1}{2}} = \gamma T^2$, where γ is a positive constant. U and V fluctuate simultaneously, and for fixed T the specific heat C is proportional to $V^{\frac{1}{2}}$. Show that the two terms of which $(\delta U)^2$ is the sum are always equal.

12.2 K is a homogeneous open system of one constituent. How must the previous generic expression for σ be modified?

12.3 T and n fluctuate simultaneously in the system of the preceding problem. Show that

$$(\delta n)^2 = kT(\partial\mu/\partial n)^{-1}.$$

12.4 If the system of the preceding problem is an ideal gas, show that the relative fluctuation of n is

$$\delta n/n = (k/Rn)^{\frac{1}{2}}.$$

12.5 U and n can fluctuate in a system consisting of an ideal gas whose specific heat is constant, i.e. $U = CT$. Show that

$$(\delta U)^2 = (1 + C/nR)kCT^2.$$

LECTURE 13

Microscopic Structure and Statistical Thermodynamics

Now that we have, hopefully, gained some insight into the phenomenological aspects of thermodynamics it will not come amiss to return to the remarks of the first lecture. The connecting link is the theory of fluctuations which was devised to give a quantitative account of the lack of quiescence of thermodynamic systems in equilibrium. What is more natural than to go further by enquiring whether this lack of quiescence which is observed on a phenomenological scale is perhaps evidence of motions of microscopic parts of which the system is made up. In other words, we abandon the pretence that 'we do not know' whether matter is continuous or discrete in structure and definitely commit ourselves to the atomic view. This means that, whatever macroscopic appearances may be, matter is taken in fact to be made up of individual particles (atoms, molecules, and the like), the number of particles in a given piece of matter of everyday dimensions being very large indeed. This brings with it that a system K which is macroscopically a thermodynamic system is microscopically a mechanical system which has an enormous number of degrees of freedom. Its detailed dynamical history is governed by the equations of motion, be they Newtonian or relativistic, classical or quantum-mechanical. It is, however, clearly out of the question to solve these equations, partly as a matter of practical realities and partly as a matter of principle, to the extent that a solution of the equations requires a knowledge of initial conditions and these will not be known in sufficient detail. We therefore have to be content with calculating certain averages, which are, generally speaking, just the counterparts of those quantities which enter into the purely phenomenological description of K. On this view, then, the thermodynamic behaviour of systems is, so to speak, the outcome of the detailed dynamical behaviour of its microscopic parts. Of course, if we can devise a successful theory which is a concrete realization of this view we shall have gained much more than a mere interpretation. This is because in order to be able to write down specific microscopic equations of motion we have to know all the forces which operate within the system, and the wealth of information so injected into the theory entails a corresponding richness of detail in the results which it provides. In particular, the values of quantities such as the bulk parameters a and κ, the specific heats, virial coefficients and the like, can be calculated, whereas in the phenomenological theory such quantities appeared merely in a formal way and the values of each of them had to be separately obtained by experiment.

The new theory which we are seeking to formulate is called *statistical thermodynamics* since it operates in terms of averages taken over those detailed dynamical histories which are consistent with whatever macroscopic information is at hand. As a cornerstone of our attempt to construct such a theory we adopt the definitive view that the phenomenological laws are to be regarded as primary laws whose validity remains unquestioned, the statistical theory then being the simplest mechanical model which will be consistent with them and, in a sense, account for them. It goes without saying that in this way we abandon any idea of achieving an *a priori* 'explanation' of the phenomenological behaviour of systems K to the extent that, given the equations of motion, one could state with certainty that the macroscopic behaviour of K *must* be such and such. The correctness of the theory must finally be tested by experiment, for example by comparing calculated values of specific heats, of compressibilities, and so on, with their measured values. Still, in the last resort one has surely much the same situation however the development of the theory is undertaken [*Note* 46]. At any rate, we are here at liberty to impose, where required, merely sufficient, rather than necessary, conditions, as long as these are consistent with the overriding requirement that the generic content of the phenomenological theory remains, so to speak, embedded in the resulting formalism.

In line with this general view we now lay down from the start that thermodynamic quantities correspond to ensemble averages rather than to temporal averages, which, as we already know, cannot be calculated. In this way we circumvent the need to concern ourselves with the question of their mutual equivalence (the somewhat intractable 'ergodic problem'). The meaning of a representative ensemble $\mathscr{K}$ of a given system K is precisely that which we attached to it in the previous lecture. All members of $\mathscr{K}$ have the same generic structure and all are prepared consistently with whatever information about K is available. This information is now provided by specifying (i) the phenomenological state $\mathfrak{S}$ of K, (ii) whether K is open or closed, (iii) whether K is in diathermic equilibrium or is (adiabatically) isolated, (iv) the microscopic structure of K, which includes all the forces operating within it. Here only the detailed microscopic description of K has not previously entered into our discussion and we must briefly turn our attention to it.

Regarded as a dynamical system, K has an enormous number r of degrees of freedom. For example, if K consists of N like particles each of which has i 'internal' degrees of freedom, $r = (i + 3)N$. One coordinate q_k of some kind is associated with each degree of freedom of K, so that there are r such coordinates $q_1, q_2, \ldots, q_r$ ($\equiv q$, for short). Their values change in the course of time, so that we have the r 'velocities' $\dot{q}$ ($\equiv \dot{q}_1, \dot{q}_2, \ldots, \dot{q}_r$), dots denoting differentiation with respect to the time t. (Of course, in saying that we are implicitly restricting ourselves for the time being to the classical, i.e. non-quantal, description of K.) The $2r$ variables q, $\dot{q}$ are independent of each other in the sense that their values can be freely specified at any instant. By contrast, given the values of the q and the $\dot{q}$, those of the 'accelerations' $\ddot{q}$ are already provided by the equations of motion $\dot{p}_k = \partial L/\partial q_k$, $(k = 1, \ldots, r)$, where the 'momenta' p_k are defined by $p_k = \partial L/\partial \dot{q}_k$, and $L(\dot{q}, q)$ is a function which can be written down once the structure of K is explicitly specified [*Note* 47]. Although it is by no means mandatory to do so, it is preferable for purely formal reasons to use in place of the $\dot{q}$ and q the $2r$ variables p and q. In place of $L(\dot{q}, q)$ we then have the functions $H(p, q)$—the *Hamiltonian* of K—in terms of which the equations of motion take the convenient form $\dot{q}_k = \partial H/\partial p_k$, $\dot{p}_k = -\partial H/\partial q_k$ [*Note* 48]. H does not depend explicitly upon the time since the forces, whether internal or external, do not. Therefore H is simply the constant total energy of the system, expressed as a function of the conjugate variables p and q. It is thus just the energy function which contains all the available information concerning the micro-structure of K.

Now, at any given time t the coordinates q and the momenta p have certain values, and this set of $2r$ values is called the *phase* (or *microstate*) of K at time t. (This use of the term 'phase' has nothing at all to do with its previous connotation.) Each member of the ensemble $\mathscr{K}$ is in its own phase and—since we are interested in statistical results—we now ask a question very reminiscent of one which arose in the theory of fluctuations. It is this: what is the probability that the phase of a randomly selected member of $\mathscr{K}$ will be in the range between (p, q) and $(p + dp, q + dq)$? This probability will be proportional to dp and dq and we write it as $h^{-r}\varphi\, dp_1 dq_1 dp_2 \ldots dq_r$, or $\varphi\, d\Gamma$ for short, $d\Gamma$ being the element of *extension in phase*. The factor h^{-r} is inserted to ensure that φ is dimensionless, h being an as yet unspecified constant of the dimensions of action, i.e. ml^2t^{-1}. φ, which is called the *probability in phase*, is a function of the dynamical variables p, q and—since it must reflect the state $\mathfrak{S}$ of K—also of the thermodynamic coordinates $x_1, x_2, \ldots, x_{n-1}, T$.

Whatever the explicit form of φ may happen to be we must of course have

$$\int \varphi\, d\Gamma = 1,$$

for this identity says no more than that any member of $\mathscr{K}$ must be in some phase. If D is any phase function, i.e. function of the dynamical variables p, q (whatever other variables it may depend on) its ensemble mean is, by definition,

$$\bar{D} = \int D\, \varphi\, d\Gamma,$$

and every appropriate thermodynamic function is an ensemble mean of this kind.

K being in equilibrium, it goes without saying that φ cannot depend explicitly upon the time: $\partial\varphi/\partial t = 0$. It will deepen our understanding to contemplate this conclusion in the light of the notion of the preparation of $\mathscr{K}$. To this end, let K, by way of example, consist of N like particles contained in an isolated enclosure of fixed volume V, the energy of K being not greater than E. With this information at hand we would seem to have no option but to prepare $\mathscr{K}$ by inserting N particles into each of ν enclosures, each of volume V, with all values of the p, q corresponding to energies $\leqslant E$ equally represented. In other words, we shall have $\varphi = \varphi_0$ (= constant) for values of the p, q such that $H \leqslant E$, whilst $\varphi = 0$ for all other values of the p, q. (It should be borne in mind that $H = \infty$ outside the enclosure, see Problem **14.1**.) The ensemble so defined is sometimes called the *quasi-uniform ensemble*. The point at issue is now this: that the constancy of φ at time $t = 0$ must be consistent with the vanishing of $\partial\varphi/\partial t$. In fact, it is not difficult to show [*Note* 49] that, always,

$$\partial\varphi/\partial t + \Sigma\, [(\partial\varphi/\partial q_k)(\partial H/\partial p_k) - (\partial\varphi/\partial p_k)(\partial H/\partial q_k)] = 0$$

as a consequence of the equations of motion *(Liouville's Theorem)*. Consequently if φ = constant at time $t = 0$, $\partial\varphi/\partial t = 0$ then, and therefore at all subsequent times. This shows that the equations of motion in themselves do not imply any tendency for the phase points to prefer one part of Γ over another [*Note* 50].

Problems

13.1 A system K consists of a polyatomic gas, each molecule consisting of s structureless atoms. There is no interaction between different molecules, and there are N of these. What is the value of r when (i) the molecules are rigid, (ii) the atoms within the molecules are elastically bound?

13.2 A system K consists of N atoms all alike, contained in an enclosure of volume V. Each possesses a mass m and an electric charge e. K is situated in an external electric field.

Disregarding (somewhat artificially) the mutual repulsion between pairs of atoms, write down the Hamiltonian of K.

13.3 The following information is given about an isolated system K: (i) the energy is not greater than E, (ii) its volume is V, (iii) there are N structureless particles of mass m between which there are no interactions. K is evidently represented again by the quasi-uniform ensemble. Find φ_0 in terms of E and V. {In order to evaluate any integral over Γ when the integrand is a function $f(r)$ of $r = (\Sigma\, p_k^2)^{\frac{1}{2}}$ alone, introduce 'spherical polar coordinates' $p_1 = r \cos \alpha$, $p_2 = r \sin \alpha \cos \beta$, $p_3 = r \sin \alpha \sin \beta \cos \gamma$, and so on, and integrate over the angle variables. Then $\int f(r) dp_1 \ldots dp_n = [n\pi^{\frac{1}{2}n}/(\frac{1}{2}n)!] \int f(r) r^{n-1} dr$.}

13.4 Find the energy U of the system of the preceding problem, granted that $U = \overline{H}$.

13.5 Continuing the previous problem, determine the entropy S, granted that $S = -k\,\overline{\ln \varphi}$, where k is a constant. In the phenomenological theory $T^{-1} = \partial S(U, V)/\partial U$. Let a function T be here defined by this relation. Show that $U = \frac{3}{2}NkT$.

13.6 We noted during the lecture that the use of canonical variables p, q is not mandatory. On the other hand if we use the $\dot{q}$ and q instead we can no longer take $d\Gamma$ = constant $\times$ $d\dot{q}_1 dq_1 d\dot{q}_2 \ldots dq_n$ since under an arbitrary transformation $q_k = f_k(q_1^*, \ldots, q_n^*)$ the form of $d\Gamma$ is not preserved, i.e. in general one does not end up with $d\Gamma$ = constant $\times$ $d\dot{q}_1^* dq_1^* \ldots dq_n^*$. Can you suggest what form $d\Gamma$ should take?

LECTURE 14

The Canonical Ensemble

We recall that in our attempt to construct a formalism of statistical thermodynamics we intend to be guided by the overriding requirement that the phenomenological theory—whose validity remains unquestioned—must necessarily be embedded within it. Our starting point will therefore naturally be the central relation $T\delta S = \delta U + \delta W$ which is, so to say, the distilled essence of the Zeroth, First and Second Laws in the context of reversible transitions. To translate this relation into the language of ensemble averages requires that we should construct an ensemble appropriate to the situation now envisaged which is evidently that of a system K in diathermic equilibrium with its surroundings, its volume V, or more generally each of its d-coordinates, having a prescribed value. Such an ensemble is called *canonical* or *grand canonical* according as K is closed or open. For the moment we shall consider the former alone.

We begin with the observation that $\triangle U$ is the work done on K when it interacts only mechanically with its surroundings. In any mechanical interpretation U must therefore be the (conserved) mechanical energy of K, i.e. the value of its Hamiltonian H. It is this value which is obtained when the energy of any member K_a of any given ensemble is measured since this measurement, we recall, is made after K_a has been effectively isolated. We conclude that U is the ensemble mean of H:

$$U = \overline{H} = \int H\,\varphi\,d\Gamma.$$

H depends of course not only on the p and q but also on the phenomenological d-coordinates $x_1, \ldots, x_{n-1}$ since their values reflect the macroscopic structure of K. An analogous situation prevails with regard to δW. The work done on a mechanical system in a variation of its constraints is exactly balanced by the change in value of its Hamiltonian, provided the variation occurs at an infinitesimal rate. This condition is just satisfied in a reversible, i.e. quasi-static, transition. Therefore δW—which is again an ensemble average—is transcribed as follows:

$$\delta W = -\overline{\delta H} = -\int(\delta H)\varphi\,d\Gamma.$$

Next, S will be the ensemble mean of some function w. Here we have a different situation to the extent that in the nature of things we cannot ascribe any sort of instantaneous entropy to a purely mechanical system. In other words, w cannot be some function defined (at time t) on each member of $\mathscr{K}$ separately but must, on the contrary, characterize $\mathscr{K}$ as a whole. Since φ does just that, no matter what the particular nature of the physical situation may be, we are naturally led to assume that w is a universal function of φ alone. This assumption is made yet more plausible by the reflection that the phenomenological entropy entered into the theory prior to any other thermodynamic function having been defined. Since now $S = \overline{w} = \int \varphi\, w(\varphi) d\Gamma$, the relation $T\delta S - \delta U - \delta W = 0$ becomes

$$\int [(\varphi w)' - H/T]\delta\varphi\, d\Gamma = 0,$$

primes denoting differentiation with respect to the argument.

Now recall that $\int \delta\varphi\, d\Gamma$ must vanish for any variation $\delta x_1, \ldots, \delta T$ of the thermodynamic coordinates. The vanishing of the previous integral is therefore assured if—in line with our declared policy—we make the sufficient assumption that the expression in square brackets is a 'constant', $-\Lambda$, say, i.e. is independent of the p, q. Then $(\varphi w)' = H/T - \Lambda$, which shows that φ is a function of $H/T - \Lambda$ alone [*Note* 51]. Now $\int \delta\varphi\, d\Gamma = 0 = \int \varphi' \delta(H/T - \Lambda) d\Gamma$, or

$$\overline{(\varphi'/\varphi)\delta(H/T - \Lambda)} = 0.$$

However, the formalism cannot contain any differential relation other than $T\delta S - \delta U - \delta W = 0$ or some concomitant of it. It is difficult—to say the least—to see how this can be reconciled with the existence of the relation just obtained other than by assuming that $\varphi'/\varphi = \text{constant} = \kappa$, say; for then we are left with a relation containing no ensemble means of a kind which cannot at once be appropriately interpreted. Thus we have now $\overline{\delta(H/T - \Lambda)} = 0$, or

$$T^{-1}\overline{\delta H} - T^{-2}\overline{H}\,\delta T - \delta\Lambda = -T^{-1}\delta W - T^{-2}U\delta T - \delta\Lambda = \delta(F/T) - \delta\Lambda = 0.$$

We conclude that $\Lambda = F/T$ to within an arbitrary additive constant which may be omitted because phenomenologically F is in any event defined only to within an arbitrary additive linear function of T. Then $\varphi = a \exp[\kappa(H - F)/T]$, where a is a numerical constant, but since φ surely cannot go to infinity as H goes to infinity—assuming here that H can do so—κ must be negative and we write $-1/k$ for it ($k > 0$). k is generally known as *Boltzmann's constant*. Thus finally

$$\varphi = a \exp[(F - H)/kT].$$

Taking the ensemble mean of $\ln \varphi$ it further follows that [*Note* 52]

$$S = \overline{w} = -k\,\overline{\ln(\varphi/a)}.$$

The identity $\int \varphi\, d\Gamma = 1$ may now be written as

$$e^{-F/kT} = a \int e^{-H/kT}\, d\Gamma.$$

To avoid any ambiguity in future, we shall understand the Helmholtz potential to be the function F calculated from this relation. The integral on the right is called the *canonical partition function*, and we write Z for it, so that

$$F = -kT \ln Z.$$

Thus, once we are given the function H which defines the microscopic structure of K we can immediately determine the function F which characterizes the macroscopic (i.e. thermodynamic) structure of K, even if the actual calculations required may be very intricate indeed. In particular, if the volume V is the only d-coordinate of K,

$$P = -\partial F/\partial V = kT\partial \ln Z/\partial V, \qquad S = -\partial F/\partial T = k\partial(T \ln Z)/\partial T,$$

$$U = F + TS = kT^2 \partial \ln Z/\partial T,$$

the first of which is the equation of state; whilst in the presence of several d-coordinates we have, more generally, $P_i = kT\, \partial \ln Z/\partial x_i$.

As regards the constant a which occurs in Z, this might very well depend upon the numbers which enter into the number of degrees of freedom of K. These have hitherto been left understood since for a given canonical ensemble they are of course fixed. For the moment we merely observe that an appeal to the extensivity of F [*Note* 53] would lead to the conclusion that $a = (N/\gamma)^{-N}$ where γ is a numerical constant. However, we prefer to determine a in the next lecture by quite different means.

The first of some general remarks on the canonical ensemble and a few generic results based on it concerns the fact that φ manifestly depends on the dynamical variables only through their occurrence in H. This at once leads to the satisfactory conclusion that the form of φ is not in conflict with Liouville's Theorem.

Next, the only independent extensive quantity which is not fixed from the start is here the energy. Its fluctuation δU is, as usual, given by $(\delta U)^2 = \overline{(H-U)^2} = \overline{H^2} - U^2$. Now $\overline{H^2} = \int H^2 \varphi\, d\Gamma = ae^{\beta F} \int H^2 e^{-\beta H} d\Gamma = Z^{-1} \partial^2 Z/\partial\beta^2$, by inspection, with $\beta = 1/kT$. Therefore, since $U = -Z^{-1}\partial Z/\partial\beta$, we have quite generally $(\delta U)^2 = \partial^2 \ln Z/\partial\beta^2 = -\partial U/\partial\beta = kT^2 C$, a result of precisely the same form as that obtained in the twelfth lecture.

General manipulations of an analogous kind also allow us to calculate the averages $X_{ij} = \overline{\chi_i\, \partial H/\partial\chi_j}$, where χ_i and χ_j are any two of the dynamical variables, subject only to the condition that H becomes infinite at both ends of the range of values which χ_j can assume. We have $X_{ij} = \int \chi_i\, (\partial H/\partial\chi_j)\varphi d\Gamma = -kT \int \chi_i\, (\partial\varphi/\partial\chi_j) d\Gamma$. Integrating this by parts with respect to χ_j, the integrated part vanishes since $\varphi = 0$ at the limits of integration, bearing in mind that we required H to be infinite there. (It is also assumed that H depends continuously on χ_j.) We are left with

$$X_{ij} = kT \int \varphi(\partial\chi_i/\partial\chi_j) d\Gamma = \delta_{ij}\, kT,$$

($\delta_{ij} = 1$ or 0 according as $i = j$ or $i \neq j$). This result is the *general theorem of equipartition*. In particular, if χ is any dynamical variable which enters into H solely through an additive term $\frac{1}{2}g\chi^2$, where g may depend on the other variables, then there is a corresponding contribution $\frac{1}{2}kT$ to U *(equipartition of energy)*.

Finally, suppose that K consists of N particles of mass m. The internal motions of each particle are to be referred to its centre of mass whose velocity, relative to cartesian axes, is $\mathbf{u}$ ($= \mathbf{p}/m$ [*Note* 54]). To the centre-of-mass motion of a particular particle there then corresponds an additive term $\frac{1}{2}m|\mathbf{u}|^2$ in H. Integrating over all the variables other than $\mathbf{p}$, $2r - 3$ of them, we conclude that the probability of the (centre of mass of the) particle having a velocity in the range ($\mathbf{u}$, $\mathbf{u} + d\mathbf{u}$) is proportional to $\exp(-\frac{1}{2}m|\mathbf{u}|^2/kT)d\mathbf{u}$. Since all the particles are equivalent we thus have *Maxwell's distribution law of molecular velocities*: the number δN of particles whose velocities lie in the range ($\mathbf{u}$, $\mathbf{u} + \delta\mathbf{u}$) is given by

$$\delta N = N(2\pi mkT)^{\frac{3}{2}} \exp[-\tfrac{1}{2}m(u_x^2 + u_y^2 + u_z^2)/kT]\delta u_x \delta u_y \delta u_z,$$

the initial factor on the right being determined by the condition that $\int \delta N = N$. This result is evidently quite unaffected by the presence of interactions of the molecules with each other or with external fields.

Problems

14.1 A system consists of N structureless particles of mass m contained in a rigid box of volume V. What is the Hamiltonian of K?

14.2 Show that for the system of the preceding problem

$$Z = a[(2\pi m h^{-2} kT)^{\frac{3}{2}} V]^N,$$

and hence determine its equation of state as well as expressions for U and C.

14.3 Show that according to the theory so far developed the magnetic susceptibility of every completely classical system is zero.

14.4 Suppose that a gaseous system K consists of N rigid particles each of which somehow has a given magnetic moment of magnitude μ. Let the mutual interaction between these be disregarded, i.e. the gas is very dilute. (i) If the gas is situated in a homogeneous magnetic field of magnitude B show that the partition function is $Z = Z_0 (y^{-1} \sinh y)^N$, where Z_0 is the partition function when $B = 0$ and $y = \mu B/kT$. (ii) Deduce that the magnetic energy of K is given by $U_m = -N\mu B(\coth y - 1/y)$. (iii) Infer that when B is sufficiently small (in fact, when $y \ll 1$) the susceptibility of K is given by $\chi = N\mu^2/3kT$.

14.5 In Problem **14.1**, suppose that the temperature is so high that the mean particle speed is comparable with the speed of light c. Is it true that the equation of state is still $PV = NkT$?

14.6 Use the theorem of equipartition to show that the specific heat C of the relativistic ideal gas (in the sense of the preceding problem) is given by $C = \frac{3}{2}Nk$ when $T \ll mc^2/k$ and $C = 3Nk$ when $T \gg mc^2/k$.

14.7 Occasionally the theorem of equipartition has been used in the context of the linear harmonic oscillator perturbed by a sufficiently small term εq^3 ($\varepsilon > 0$), i.e. $H = \frac{1}{2}(p^2 + q^2) + \varepsilon q^3$, it being stated that $\overline{q\partial H/\partial q} = \overline{q^2} + 3\varepsilon\overline{q^3} = kT$. What is wrong with this?

LECTURE 15

The Grand Canonical Ensemble. The Constant α. The Constant-Pressure Ensemble

In the phenomenological theory we first confined ourselves to closed systems and later went on to generalize the formalism so as to allow for the possibility of the system under consideration being able to exchange matter with its surroundings. In an analogous fashion we must now enlarge the statistical theory, that is to say, we must contemplate an ensemble $\mathscr{K}$, called the *grand canonical ensemble*, which represents a system K capable of exchanging not only energy with its surroundings but particles as well. Any randomly selected member will now have a certain phase and will contain N_1, N_2, ... particles of the various constituents C_1, C_2, ..., K being supposed homogeneous. The phenomenological mole number n_j which specifies the amount of C_j in K is to be identified with the ensemble mean $\bar{N}_j/L$, where L *(Avogadro's number)* is the fixed number of particles which makes up one mole of any substance. As a matter of convenience we henceforth understand μ_j to be the chemical potential per particle of C_j, which is $1/L$ times the quantity previously denoted by μ_j.

For the time being it will suffice to suppose that K is composed of a single constituent. Then we are concerned with the probability $\varphi_N(p,q;x)d\Gamma_N$ that a randomly selected member of $\mathscr{K}$ will be found to contain N particles and to have its phase point lying within the element $d\Gamma_N$ of a 2r-dimensional phase space. To determine the form of the grand canonical probability in phase φ_N we proceed in complete analogy with the argument pursued in the last lecture, that is to say, whenever required we make the simplest sufficient assumption which will lead us to an expression for φ_N such that the central phenomenological relation $T\delta S = \delta U + \Sigma\, P_k\, \delta x_k - \mu\delta\bar{N}$ is necessarily embedded within it. As before, the validity of the final result must then ultimately be tested by experiment.

If D is any phase function (which may depend on N) its phenomenological analogue is the ensemble mean

$$\bar{D} = \sum_N \int D\varphi_N\, d\Gamma_N.$$

The quantity $\bar{N} = \sum_N N \int \varphi_N\, d\Gamma_N$ which we have already talked about is one example. If H_N is the Hamiltonian of an ensemble member K_N containing just N particles, then we have, effectively as before, $U = \bar{H}_N$ and $\Sigma\, P_k\, \delta x_k = -\overline{\delta H_N}$. As for S, we again take it to be the ensemble mean of a function w of φ_N alone. In this way we are led to the equation

$$\sum_N \int [(\varphi_N w)' - (H_N - \mu N)/T]\delta\varphi_N d\Gamma_N = 0.$$

We can continue almost word for word the argument relevant to the canonical ensemble. Thus, we satisfy the last equation by assuming that the expression in square brackets is a function, $-\Lambda$ say, which does not depend on the dynamical variables, bearing in mind that $\Sigma \int \delta\varphi_N d\Gamma_N = 0$. We conclude that φ_N is a function of $(H_N - \mu N)/T - \Lambda$ alone. Again we are virtually forced to assume that φ_N'/φ_N is a constant, so that we are left with the equation $\delta[(H_N - \mu N)/T - \Lambda] = 0$. This at once becomes

$$T^{-1}\Sigma P_k\,\delta x_k + T^{-2}\,U\,\delta T + \bar{N}\,\delta(\mu/T) + \delta\Lambda \equiv \delta(X/T + \Lambda) = 0.$$

We conclude that $-T\Lambda$ is to be identified with the grand potential X. Thus finally

$$\varphi_N = a_N^* \exp[(-X + \mu N - H_N)/kT],$$

where a_N^* is a constant. Also, $w = -k \ln(\varphi_N/a_N^*)$, and

$$e^{X/kT} = \Sigma\, a_N^* \int e^{(\mu N - H_N)/kT} d\Gamma_N.$$

The expression on the right is the *grand canonical partition function* which we denote by Z. Then

$$X = kT \ln Z,$$

so that, given only H_N, we can compute Z, and this in turn gives

$$S = k\partial(T \ln Z)/\partial T, \qquad P_k = kT\partial \ln Z/\partial x_k, \qquad \bar{N} = kT\partial \ln Z/\partial\mu,$$
$$U = kT(\mu\partial \ln Z/\partial\mu + T\partial \ln Z/\partial T).$$

Formally H_N defines the canonical partition function Z_N of a system of exactly N particles. If we now also introduce the abbreviation λ for the *(absolute) activity* $\exp(\mu/kT)$, we can write elegantly

$$Z = \Sigma\, a_N Z_N \lambda^N,$$

with $a_N = a_N^*/a_N$. This result reflects the nature of the grand canonical ensemble in as far as it is a weighted collection of canonical ensembles, $\int \varphi_N d\Gamma_N = a_N Z_N \lambda^N$ being the probability that a randomly selected member of $\mathscr{K}$ belongs to the canonical sub-ensemble all members of which contain just N particles.

Now, the relative fluctuation $\delta N/\bar{N}$ of the particle number of an ideal gas is exactly $\bar{N}^{-\frac{1}{2}}$ [*Note* 55]. As regards orders of magnitude, this result will also hold for non-ideal systems under conditions in which their behaviour is not pathological [*Note* 56]. Therefore, when $\bar{N}$ is sufficiently large (that is, in the thermodynamic limit) $\mathscr{K}$ represents to all intents and purposes a system containing just $\bar{N}$ particles, so that its Helmholtz function F can be calculated equally well from $Z_{\bar{N}}$ or from Z. In short, asymptotically we must have $-kT \ln Z_{\bar{N}} \sim \bar{N}\mu - X = \bar{N}\mu - kT \ln Z$. Since $Z = \Sigma\, a_N Z_N \lambda^N$, this is possible only if $a_N Z_N \lambda^N$, regarded as a function of N, has a steep maximum at $N = \bar{N}$. The condition $\partial(a_{\bar{N}} Z_{\bar{N}} \lambda^{\bar{N}})/\partial\bar{N} = 0$ must therefore be satisfied, that is, we must have $\partial \ln a_{\bar{N}}/\partial\bar{N} + \partial \ln Z_{\bar{N}}/\partial\bar{N} + \ln\lambda = 0$. However, If F be calculated from $Z_{\bar{N}}$, $\mu = \partial F/\partial\bar{N}$, or $\partial \ln Z_{\bar{N}}/\partial\bar{N} + \ln\lambda \sim 0$, and we are left with $\partial \ln a_{\bar{N}}/\partial\bar{N} \sim 0$. On the other hand, the number $\omega_{\bar{N}}$ of terms of the sum for Z which are comparable in magnitude with the largest term is of the order of $\bar{N}^{\frac{1}{2}}$ and we may disregard all other terms in the limit $\bar{N} \to \infty$. Therefore $\ln Z \sim \ln(\omega_{\bar{N}}\, a_{\bar{N}}) + \ln Z_{\bar{N}} + \bar{N}\ln\lambda$ and comparison with the earlier asymptotic relation between Z_N and $Z_{\bar{N}}$ implies that $\ln a_{\bar{N}} \sim -\ln\omega_{\bar{N}} \sim -\frac{1}{2}\ln\bar{N}$, which is consistent with the condition $\partial \ln a_{\bar{N}}/\partial\bar{N} \sim 0$.

To determine the constants a_N^* we first recall (supposing there to be only the single d-coordinate V) that $\partial P(V, T, \mu)/\partial V = 0$ always, which implies that $\ln Z$ must depend linearly upon V. For further progress it suffices to consider an ideal gas [*Note* 57]. For this [cf. Problem **14.2**] $Z_N = a_N(\tau V)^N$, where τ is a function of T alone. Therefore $Z = \Sigma a_N^*(\lambda\tau V)^N$ and this series must represent an exponential function of V, as we have just seen. It follows that $a_N^* = a_0^* b^N/N!$. The value of the fixed constant a_0^* is immaterial and we take $a_0^* = 1$, whilst b can be absorbed in the unspecified constant h contained in $d\Gamma_N$. We are thus finally left with the exact result $a_N^* = 1/N!$. From this in turn, $\ln a_N \sim -\ln(N!) + \frac{1}{2}\ln N$, that is, $\ln a_N \sim -\ln(N!)$, since $\ln(N!) \sim N\ln N$. The mutual thermodynamic equivalence of the canonical and grand canonical ensembles is therefore safeguarded by taking $a_N = 1/N!$. Then, furthermore, $\ln a_N \sim -N\ln N + N$ which is in harmony with the conclusion that $\ln a_N \sim -N\ln N + N\ln\gamma$ to which we already alluded in the previous lecture.

The generalization to several constituents $C_1, \ldots, C_z$ is not difficult to achieve and it suffices to write down the final result. If $H_{N_1 \ldots N_z}$ is the Hamiltonian of a system definitely containing N_1 particles of C_1, N_2 particles of C_2, and so on, define

$$Z_{N_1 \ldots N_z} = (N_1!N_2! \ldots N_z!)^{-1} \int \exp(-H_{N_1 \ldots N_z}/kT)d\Gamma_{N_1 \ldots N_z}.$$

Then

$$Z = \sum_{N_1} \ldots \sum_{N_z} \lambda_1^{N1}\lambda_2^{N2} \ldots \lambda_z^{Nz} Z_{N_1 \ldots N_z},$$

and $X = kT\ln Z$; from which the various thermodynamic functions are obtained in the standard way by differentiation.

It is instructive to return to the need to go over from the canonical to the grand canonical ensemble from another point of view. Clearly the canonical ensemble was no longer appropriate as soon as we changed the generic physical conditions under which it was defined, or in other words, as soon as we asked about the fluctuations of N once the system was no longer closed. An exactly analogous situation prevails if we enquire into the volume fluctuations of a system for which P, T and N have given values. Again we must construct the relevant ensemble $\mathscr{K}$—the *constant-pressure ensemble*. Its probability in phase $\varphi(p, q, V; P, T)$ gives the probability $V_0^{-1}\varphi\, d\Gamma\, dV$ that the volume of a randomly selected member of $\mathscr{K}$ lies in the range $(V, V + dV)$ whilst its phase lies in $d\Gamma$. (V_0 is an unspecified fixed volume which makes φ dimensionless.) Going through the usual argument once again, it turns out that

$$\varphi = (1/N!)\exp[(G - PV - H)/kT],$$

and therefore

$$G = -kT\ln Z, \qquad Z = (1/N!)V_0^{-1}\int \exp[-(PV + H)/kT]d\Gamma\, dV.$$

Not surprisingly $G(P, T)$ here appears in place of $F(V, T)$. Except in as far as V is a continuous variable, these equations formally resemble those relating to the grand canonical ensemble very closely indeed. They are, moreover, especially well suited to exemplifying how one may return from our present formalism to the phenomenological theory of fluctuations. Thus, considering volume fluctuations, the probability $\psi(V)dV$ that the volume lies in the range $(V, V + dV)$ is, by inspection, $V_0^{-1}\, dV \int \varphi\, d\Gamma = V_0^{-1}\, dV\, Z\,(V, T) \times \exp[(G - PV)/kT]$, where $Z(V, T)$ is formally the canonical partition function. Writing $Z = \exp[-F(V, T)/kT]$, we therefore have $\psi(V) = V_0^{-1}\exp\{[G(P, T) - \hat{G}(V, T)]/kT\}$, where $\hat{G}$ is the Gibbs potential formally regarded as a function of V and T. This result is evidently in complete harmony with the earlier equation $\psi(V) = \kappa\, e^{-\sigma/k}$, with $\sigma = \triangle G/T$, granted that in the exponent k is indeed identified with Boltzmann's constant.

Problems

15.1 A gaseous system K consists of $\bar{N}_1$ particles of one kind and $\bar{N}_2$ particles of another, contained at temperature T in a box of volume V. The particles do not interact with each other. Determine the grand canonical partition function and hence the entropy of K as a function of V, T, $\bar{N}_1$, $\bar{N}_2$.

15.2 The two gases of the preceding problem are contained at the same pressure P and at temperature T in separate boxes in diathermic contact. (i) Compute the entropy of this system. (ii) Show that when the partition between the two boxes is removed the resulting process of diffusion is accompanied by an increase of entropy

$$\triangle S = \bar{N}_1 \ln (1 + \bar{N}_2/\bar{N}_1) + \bar{N}_2 \ln (1 + \bar{N}_1/\bar{N}_2) > 0.$$

(iii) The result just obtained is independent of the nature of the two gases and will hold for instance when the one is hydrogen and the other deuterium. However, when both are hydrogen, say, $\triangle S$ must surely be zero, which seems to contradict the general result. What can you say about this at first sight strange situation?

15.3 Implicitly or explicitly the following mathematical question often arises in the theory: evaluate the integral $J = \int_0^\infty e^{-f(x)} dx$ given that $f(x)$ has one steep minimum between 0 and ∞, at $x = x_0$, say. (i) Find an approximation to J by expanding $f(x)$ as a power series about the point $x = x_0$. (ii) Apply your result to find an approximation to $N! = \int_0^\infty x^N e^{-x} dx$ for large N.

15.4 Use the constant-pressure ensemble to show that the root mean square volume fluctuation is given by $\delta V = (-kT\, \partial \bar{V}/\partial P)^{\frac{1}{2}}$.

15.5 Let $g(N)$ be a function which has a steep maximum, at $N = N_0$ ($\gg 1$), say. During the lecture we argued that $\sum_{N=0}^{\infty} g(N)$ could be sufficiently approximated by $\gamma\, N_0^{\frac{1}{2}}\, g(N_0)$. Examine this statement in the case where $g(N) = y^N/N!$, y = constant.

15.6 Has the presence of the constant V_0 in the partition function of the constant-pressure ensemble any significant physical consequences?

LECTURE 16

The Entropy Principle. Implications of the Third Law

The time has come to remind ourselves that the content of the phenomenological theory is by no means exhausted by the central relation $T\delta S - \delta U - \delta W = 0$ which has hitherto been our sole guide towards the construction of a statistical formalism. On the contrary, we have, in the first place, not even mentioned the Entropy Principle, and to this we now turn our attention. In one respect we face here the same difficulty as previously, in as far as we are not in a position to study the approach to equilibrium. Let us recall that we helped ourselves earlier by the device of introducing pseudo-states. Thus, a system K could find itself in a pseudo-state $\mathfrak{S}^*$, distinct from but corresponding to the state $\mathfrak{S}$ defined by the given values of the thermodynamic coordinates, either by the temporary introduction of material partitions or else spontaneously as the result of a large fluctuation. The basic assumption was that—in the first case upon removal of the partitions and in the second case spontaneously—the system would of necessity undergo a change so that it would eventually end up in the state $\mathfrak{S}$. When K was, at given volume, in diathermic equilibrium with its surroundings—the situation of exclusive interest to us now—the Entropy Principle then led us to conclude that the value of the Helmholtz potential F of every pseudo-state $\mathfrak{S}^*$ corresponding to $\mathfrak{S}$ must be greater than that of $\mathfrak{S}$.

We now examine the statistical counterpart of these ideas. Evidently a pseudo-state of K—it suffices to take it to be closed—is now represented by an ensemble which is not canonically distributed. Then if F^* is the mean of $f = H - Tw$, calculated for this ensemble, F^* must always be greater than the canonical ensemble mean of f. Unless this turns out to be the case we shall reluctantly have to abandon what has already been established.

Let φ, φ^* be the canonical and non-canonical probabilities in phase respectively. The normalization conditions $\int \varphi \, d\Gamma = 1$, $\int \varphi^* \, d\Gamma = 1$ must of course be separately satisfied. Since $w = -k \ln (\varphi/a)$, we have $F^* = \int [H + kT \ln (\varphi^*/a)]\varphi^* \, d\Gamma$. It is convenient to write $\varphi^* = \varphi \, e^{\eta}$, where η is an arbitrary phase function. Then, bearing in mind that $\ln (\varphi/a) = (F - H)/kT$, we have at once $F^* = F + kT \int \eta \, e^{\eta} \varphi \, d\Gamma$. However, since $\int (e^{\eta} - 1)\varphi d\Gamma$ always vanishes because of the normalization conditions, we can write $F^* - F \equiv \triangle F = kT \int [(\eta - 1)e^{\eta} + 1] \, \varphi \, d\Gamma$. The integrand decreases monotonically from 1 to 0 as η goes from $-\infty$ to 0 and thereafter goes monotonically to ∞ as $\eta \to \infty$. In short, $\triangle F > 0$ unless the ensemble is canonical. Happily therefore the phenomenological extremal property $\triangle F \geqslant 0$ has its precise counterpart in the statistical theory.

Alas, we must not rashly jump to the conclusion that our theory is correct as it stands: only experiment can decide that. Here we very soon find ourselves in trouble. A diatomic gas such as oxygen or nitrogen is effectively ideal under 'ordinary' conditions. If the molecules were regarded as rigidly composed of two structureless atoms the Hamiltonian would simply consist of five additive quadratic terms and the theorem of equipartition of energy would immediately predict the value $\frac{5}{2}Nk$ for the specific heat C. This is just what is observed. Nevertheless we must feel very uneasy, for should we not expect the atoms to vibrate relatively to each other? We might as a first approximation, suppose the restoring forces to be linear, with the consequent addition of two further quadratic terms to H (representing the kinetic and potential energies of this motion), so that C should have the value $\frac{7}{2}Nk$. Experiment in fact shows that as the temperature increases the value of C rises towards $\frac{7}{2}Nk$ and it therefore appears that at normal temperatures and below the vibrational degree of freedom somehow becomes ineffective. Moreover, the atoms themselves are not structureless and they have internal motions. Their failure to be represented in C shows that at ordinary temperatures their internal degrees of freedom are also ineffective. Putting two and two together, we naturally ask ourselves whether it is not perhaps true that *all* degrees of freedom will fail to contribute to C as T becomes sufficiently small.

This question at once brings us to the last major component of the phenomenological theory which we have hitherto disregarded in the statistical theory, namely the Third Law. This certainly requires that C go to zero with T and so indeed confirms our suspicion quite generally. In any event, we have seen that the entropy function of any ideal gas involves V only through the additive term $Nk \ln V$, so that $\partial S/\partial V$ does not become independent of V as $T \to 0$, in conflict with the Third Law. The conclusion that our theory requires a fundamental modification after all is now inevitable.

The breakdown of our formalism evidently occurs whenever the mean energies which it predicts—be they of translation, rotation, or whatever—become sufficiently small. This observation is crucial, for it is just under these circumstances that the motions in question cannot even approximately be described by means of the equations of classical dynamics. That part of the framework of the theory which hitherto made explicit reference to them must therefore be demolished and be replaced by some quantum mechanical counterpart.

Of course we are not surprised at this turn of events, for we were rather foolish from the outset. Quite early we brought in the idea of the phase (or microstate) as a set of values of the coordinates and momenta; yet according to the Heisenberg Uncertainty Principle the values of a coordinate and its conjugate momentum cannot simultaneously be measured with arbitrary precision. What we have hitherto called a microstate is therefore meaningless. A moment's reflection leads us henceforth to understand the microstate of K to be its *state* in the basic quantum mechanical sense. At the same time the notion of the phase space has to be abandoned for there is no longer any sense in it. On the other hand, the device of the representative ensemble survives essentially unchanged [*Note* 58].

Now, previous experience with the formal aspects of the transition from classical to quantum mechanics surely leads us to suspect that we should search for a quantum mechanical observable (i.e. an hermitian operator) which can take the place of the probability in phase, this being no longer defined. Such an observable, called the statistical operator, indeed exists [*Note* 59]. However, we can avoid having to encumber ourselves explicitly with it or with the corresponding quantum-theoretical background, provided we agree to make a small concession. This is a restriction—of no great consequence to us here—on the character of those quantities whose ensemble means we might wish to compute. Thus, we require that any such observable $\hat{A}$ shall have a definite value in each of the stationary states (i.e. the eigenstates of H) in which K may be (in technical language: $\hat{A}$ must commute with H). Then, as a counterpart to the probability in phase φ, we can directly introduce the *probability of state* φ_j that a randomly selected member of $\mathscr{K}$ is in its jth stationary state (whose energy is E_j); for in such a state $\hat{A}$ will have a definite value A_j and the thermodynamic analogue of $\hat{A}$ will have the value

$$\bar{A} = \Sigma\, A_j\, \varphi_j.$$

We must not forget, by the way, that there may be several states—say g_j of them—with a given energy E_j, and then the sum will include g_j corresponding terms. One says that g_j is the *multiplicity* or *weight* of the energy level E_j.

To appreciate better what is involved in this, consider a single structureless particle in a rigid cubical box of volume V. The wave-function of the particle must vanish on the walls of the box so that according to Schrödinger's equation the energy of the particle can only have one of the discrete set of values $\varepsilon_i = (h^2/8m)\, V^{-\frac{2}{3}}\, (a_i^2 + b_i^2 + c_i^2)$, where a_i, b_i, c_i are positive integers [*Note* 60]. The number of states whose energy is less than ε is therefore the number of vertices of a rectangular unit lattice contained in one octant of the sphere of radius $(8m\, h^{-2}\, V^{\frac{2}{3}}\varepsilon)^{\frac{1}{2}}$. Granted that ε is not too small, this number is sufficiently approximated by the volume of the octant, i.e. $\frac{1}{6}\pi(8mh^{-2}\,\varepsilon)^{\frac{3}{2}}\,V$. The derivative of this gives the number of states $g(\varepsilon)d\varepsilon$ with energy in the range $(\varepsilon,\ \varepsilon + d\varepsilon)$, that is, the *density of states* $g(\varepsilon) = 4\pi mh^{-3}\, V(2m\varepsilon)^{\frac{1}{2}}$. $g(\varepsilon)$ is the analogue in the continuous limit of the single-particle weight g_j. If the particle is not structureless each state has weight ω, say, and $g(\varepsilon)$ then takes this additional factor. For a particle with intrinsic spin s and non-zero rest mass, $\omega = 2s + 1$.

When we deal with a system K containing many non-interacting particles, all of the same kind, it turns out that the single-particle density of states is all we need to know. On the other hand we must then pay due heed to the fundamental requirement of quantum mechanics that wavefunctions of K must be either symmetric or anti-symmetric under the mutual interchange of the coordinates of two particles [*Note* 61], depending on whether the particles are bosons or fermions respectively, i.e. have integral or half-odd integral spin. This entails that each single-fermion state can be occupied at most once, whereas there is no such limitation in the case of bosons.

Problems

16.1 An open system of fixed volume V is in diathermic equilibrium with its surroundings. Show that $\triangle X < 0$ unless the ensemble is grand canonical.

16.2 A two-dimensional harmonic oscillator has energy levels

$$\varepsilon(n_1, n_2) = H[(n_1 + \tfrac{1}{2})\nu_1 + (n_2 + \tfrac{1}{2})\nu_2],$$

where the n_j are positive integers or zero and the ν_j are two given frequencies. What can you say about the weight of any particular level ε?

16.3 Show that in the relativistic domain the density of states is given by

$$g(\varepsilon) = 4\pi\omega(ch)^{-3}\, V(\varepsilon^2 + 2mc^2\varepsilon)^{\frac{1}{2}}(\varepsilon + mc^2).$$

LECTURE 17

The Quantum-Statistical Canonical and Grand Canonical Ensembles

In the course of the fourteenth lecture we decided upon the specific form of the canonical probability in phase φ which characterized the ensemble $\mathscr{K}$ intended to represent a closed system K in diathermic equilibrium with its surroundings, its d-coordinates having specified values. Having now abandoned the classical description of mechanical systems, we need to find the form of the canonical probability of state φ_j which replaces φ. Inevitably we rest our case in the first instance again on the primary need to accommodate the relation $T\delta S - \delta U - \delta W = 0$. On the other hand, the improved version of the theory which we are seeking must surely contain the earlier one in some approximation and, superficially at least, a fairly close formal resemblance between the two is to be expected. What therefore is more natural than to take a shortcut: guided by our previous experience we simply prescribe the most 'obvious' form of φ_j and then verify that it satisfies our generic requirements.

Recall that φ_j is the probability that a randomly selected member of $\mathscr{K}$ will be found to be in its jth stationary state whose energy is E_j. All information about the detailed structure of K is, as before, contained in its Hamiltonian H. In the classical situation the dynamical variables entered into φ only through their occurrence in H, that is to say, φ depended on the microstates only through the values which H assumed in them. The straightforward counterpart to this feature in the quantum mechanical situation is that the probability of state should depend only on the eigenvalues of H, that is, φ_j should depend only on E_j—apart of course from the thermodynamic coordinates. Guided by the form of φ we are thus led to investigate the consequences of the assumption that

$$\varphi_j = \gamma\, e^{(F-E_j)/kT},$$

where γ is a constant; for the appearance of this strongly suggests that it will immediately lead to the relation $T\delta S - \delta U - \delta W = 0$.

To verify that it does so, observe that $\Sigma\, \varphi_j = 1$, the sum going over all states. Varying the coordinates, we get at once $\delta(F/T) + T^{-2}\, \delta T\, \Sigma E_j \varphi_j - T^{-1}\, \Sigma(\delta E_j)\, \varphi_j = 0$. Here $\Sigma\, E_j\, \varphi_j = U$, and since δE_j is the expectation value in the jth state of the change δH of the Hamiltonian [*Note* 62], $\Sigma(\delta E_j)\varphi_j = \Sigma\, \langle \delta H \rangle_j\, \varphi_j = \overline{\delta H} = -\delta W$. Thus we have $T\delta(F/T) + U\delta T/T + \delta W = 0$ which is the same as $T\delta S - \delta U - \delta W = 0$. We therefore have good reason to calculate the various thermodynamic quantities from the Helmholtz potential $F = -kT \ln Z$, granted that the partition function is now taken to be the *sum over states*

$$Z = \gamma\, \Sigma e^{-E_j/kT}.$$

Note incidentally that $S = -k\, \Sigma\, \varphi_j \ln (\varphi_j/\gamma)$.

It seems we are on the right track, the more so as a moment's reflection shows that the Entropy Principle is accommodated by φ_j just as much as it was by φ [*Note* 63]. The question which remains is whether we are now in a more favourable position as regards the Third Law, for it was the conflict between this and the consequences of the classical formalism which earlier proved to be an unsurmountable obstacle. To answer it, write E'_0 for the lowest energy level, E'_1 for the next higher level and so on, so that $Z = \gamma \Sigma g_j e^{-\beta E'_j}$, where as before, $\beta = 1/kT$. For sufficiently small values of T the numbers $\beta E'_0, \beta E'_1, \ldots$ will form a rapidly increasing sequence so that all but the first term can be disregarded altogether, i.e. $Z \sim \gamma g_0 e^{-\beta E'_0}$. Therefore, as $T \to 0$

$$\lim S = \lim \partial(kT \ln Z)/\partial T = \ln(\gamma g_0).$$

S thus indeed tends to a finite limit as $T \to 0$, a limit which is, moreover, independent of the other thermodynamic coordinates. There is therefore no longer any conflict with the Third Law. This happy situation is formally a consequence of the fact that we are now concerned with discrete rather than continuous distributions of states, appropriately understood. It is as the result of integration that previously a factor—depending on T—appeared in the partition function which was responsible for the divergence of S.

We can go on straight away to the grand canonical ensemble in very much the same way as we did in the classical case. If there is only one constituent, the grand canonical probability of state φ_{Nj} is the probability that a randomly selected member K_α of the ensemble $\mathscr{K}$ representing a system K with prescribed values of the d-coordinates and in diathermic equilibrium with its surroundings will be found to contain N particles and be in the jth stationary state (that is, the jth eigenstate of the Hamiltonian H_N of K_α). If E_{Nj} is the energy of this state,

$$\varphi_{Nj} = \gamma_N^* e^{(-X+\mu N-E_{Nj})/kT},$$

where the γ_N^* are numerical constants. As usual, the thermodynamic behaviour of K is wholly implicit in the grand potential $X = kT \ln Z$, where

$$Z = \sum_N \sum_j \gamma_N^* \lambda^N \exp(-E_{Nj}/kT).$$

At this point we go on to consider a system of N particles, all alike, under the following special conditions: first, the particles are structureless except to the extent that they may have intrinsic spin; second, they are non-interacting (in the sense that H contains no potential energy term); third, the temperature is sufficiently high in a sense which we can make more precise later on. Indeed, what we are intending to do is to deal with a one-constituent ideal gas under circumstances in which the classical description should be adequate. In this way we can kill two birds with one stone, namely, obtain the values of the constants γ_N^* and confirm at the same time that we do in fact get results equivalent to those obtained earlier.

We have to evaluate the sum $Z'_N \equiv \Sigma e^{-\beta E_{Nj}}$ over all states of K. In the absence of interactions the energy of K is the sum of the energies of the individual particles and any particular state of K whose energy is E_{Nj} can be specified by its *occupation numbers* n_{ji}, that is, the numbers $n_{j0}, n_{j1}, \ldots$ of particles which are in the single-particle states whose energies are $\varepsilon_0, \varepsilon_1, \ldots$. This is in harmony with the fact—already mentioned at the end of the previous lecture—that the wave function of every allowed state of K as a whole must be either symmetric or anti-symmetric. In the present context it is therefore the symmetrized or anti-symmetrized products of N single-particle wave functions [*Note* 64], so that all one can say is that a particle *level* ε_i (i.e., one-particle wave function with energy ε_i) occurs n_{ji} times in it. In the high temperature limit the particles will be distributed over a very large number of distinct levels so that states in which any particular level is occupied more than once can be disregarded. Now define $z = \Sigma\, e^{-\beta\varepsilon_i}$. Then, by inspection, the series for z^N contains all those terms of $Z_N' = \Sigma \exp(-\beta\, \Sigma\, n_{ji}\, \varepsilon_i)$ which have $n_{ji} < 2$ exactly $N!$ times. It does not correctly account for the other terms of Z_N', but these have already been rejected in any event since they correspond to multiple occupancy of levels. In short, in the high temperature limit $Z_N' = z^N/N!$.

We now have $Z = \Sigma\, \gamma_N^*\, Z_N'\, \lambda^N = \Sigma\, \gamma_N^*\, (\lambda z)^N/N!$. On the other hand, V is a linear factor of z because $g(\varepsilon)$ is simply proportional to V, as we have seen. If we recall that Z must be an exponential function of V, we conclude that γ_N^* must be independent of N. Its actual value has no significance and may be chosen at will. We take $\gamma_N^* = 1$, and since $\gamma_N^*/\gamma_N \sim 1$ $(N \to \infty)$ as in the classical case, we then take $\gamma_N = 1$. In short, the canonical and grand canonical partition functions are simply

$$Z_N = \sum_j e^{-\beta\varepsilon_j}, \qquad Z = \sum_N Z_N \lambda^N.$$

It remains to work z out explicitly. We have $z = \int_0^\infty g(\varepsilon)\, e^{-\beta\varepsilon}\, d\varepsilon$, since for small β it is entirely justified to replace summation by integration. Inserting for $g(\varepsilon)$ the non-relativistic expression already derived last time, we are merely left with the task of evaluating the integral $\int_0^\infty \varepsilon^{\frac{1}{2}}\, e^{-\beta\varepsilon}\, d\varepsilon$, the result being $\frac{1}{2}\sqrt{\pi}\beta^{-\frac{3}{2}}$. Finally, therefore,

$$Z_N = [\omega(2\pi m h^{-2}\, kT)^{\frac{3}{2}}\, V]^N/N!.$$

If we take the particles to be structureless, so that $\omega = 1$, we can compare this expression with that obtained in the classical theory (see the solution to Problem **14.2**). Formally they are evidently identical. However, we have now gained the additional knowledge that the previously unspecified constant h is in fact Planck's constant.

It is worth making another comment on the factor $1/N!$ which appears in φ_N and therefore ultimately in Z_N. A widely held opinion has it that the presence of this factor can only be explained on the quantum mechanical level, and that it is, as it were, arbitrarily inserted into the classical formalism. In the light of the fact that we had no difficulty in inferring that φ_N must contain just the factor $1/N!$ before we ever thought of the quantum mechanical situation surely makes the validity of this claim a little difficult to sustain, granted that the generic ensemble formalism as such be accepted. The very meaning of the numbers N_1, N_2, . . . in the grand canonical probability in phase surely implies that one has various species of particles, but that those of one particular species are in no way distinguished from each other. In other words, we can hardly admit to the notion of indistinguishability of particles as being purely quantum mechanical in character. At any rate, suppose we have an ideal gas of N structureless particles, all distinct, yet with equal masses. If we allow ourselves the view that its behaviour is represented by the subensemble $\mathscr{K}_1$ of a grand canonical ensemble such that all members of $\mathscr{K}_1$ have $N_1 = N_2 = \ldots = N_N = 1$, then $Z_{N_1 \ldots N_N}\ (= Z_{11 \ldots 1}) = (\tau V)^N$ [see the solution to problem **15.1**]. When the particles are not distinct we have to contemplate $Z_N = (\tau V)^N/N!$ instead. Thus in the first case we simply have to omit the factor $1/N!$—and this is again consistent with the quantum mechanical prescription.

Problems

17.1 A system K consists of N fixed linear harmonic oscillators. These all have the frequency ν, but they are to be regarded as distinguishable. The interaction between them allows the equilibrium of K to establish itself but is so weak that it may be otherwise disregarded. Find the classical canonical partition function of K and hence its entropy, energy and specific heat.

17.2 Treat the preceding problem using the methods of quantum statistics.

17.3 Compare in detail the results obtained in the two preceding problems.

17.4 In problem **14.4** suppose that according to the requirements of quantum mechanics the component of the magnetic moment in the direction of the field can only have one or other of the values κm, where κ is a certain constant, $m = -J, -J+1, \ldots, J$ and J is one of the numbers $\frac{1}{2}, \frac{3}{2}, \frac{5}{2}, \ldots$. Compare the magnetic part of the sum over states, with that of the classical partition function.

LECTURE 18

Semi-Classical Gaseous Systems

The distinction which we drew between formal and proper applications of phenomenological thermodynamics has a precise counterpart in the statistical theory. That the number-fluctuation in an open one-constituent system is given by $\delta N = kT(\partial^2 \ln Z/\partial\mu^2)^{\frac{1}{2}}$ is therefore an example of a formal application; proper applications on the other hand derive from explicit knowledge about the form of specific Hamiltonians. However, even the 'simplest' realistic system will have a Hamiltonian so complicated that the exact evaluation of the partition function presents almost insurmountable difficulties. We must bear in mind that the theory of real gases should embrace the phenomenon of phase transitions—liquefaction and solidification—and even qualitatively it is by no means easy to see how partition functions can in fact have the appropriate properties. This problem is one of profound interest and any attempt at its solution involves the exercise of much manipulative skill in examining the subtleties of the analytic behaviour of the grand canonical partition function in the complex λ plane. This kind of thing is clearly out of bounds here. It must therefore suffice to choose a few elementary situations which are easily tractable in some approximation.

We consider a gas consisting of N molecules, all alike, contained in a rigid box of volume V. The interactions between the molecules will depend in a complicated way upon their internal states. To be able to proceed on a simple level we therefore make the rather crude simplification that the force which the ith molecule exerts upon the jth is a function of their mutual distance r_{ij} alone and has the potential u_{ij} ($\equiv u(r_{ij})$). Here we are using non-quantal language as far as the intermolecular interactions are concerned. Granted that there is no external field, H now has the form $H_p + H_q + H_i$, H_p representing the motion of the centres of mass, H_q the intermolecular forces and H_i the internal motions and interactions. Correspondingly Z splits up into a product $Z_p Z_q Z_i$, and in turn, $F = -kT(\ln Z_p + \ln Z_q + \ln Z_i)$. To be consistent, we must suppose the conditions to be such that Z_p and Z_q can be calculated classically; in other words, we must suppose the gas to be sufficiently dilute and its temperature to be not too low. Then, in particular, $Z_p = (1/N!)[h^{-1} \int \exp(-p^2/2mkT)dp]^{3N} = l^{-3N}/N!$, where $l = (2\pi m h^{-2} kT)^{-\frac{1}{2}}$ is the *thermal wavelength* at temperature T.

The volume V enters into Z_q alone, so that Z_p and Z_i can be discarded as far as the equation of state is concerned. Now $Z_q = \int \exp(-\beta \Sigma u_{ij})dq$, where dq stands for $d\mathbf{r}_1 d\mathbf{r}_2 \ldots d\mathbf{r}_N$, with $d\mathbf{r}_j = dx_j dy_j dz_j$. We can write equivalently, with $f_{ij} = e^{-\beta u_{ij}} - 1$,

$$Z_q = \int \prod_{i<j} e^{-\beta u_{ij}}\, dq \equiv \int \prod_{i<j} (1 + f_{ij})dq.$$

If the gas were ideal all f_{ij} would vanish, and then $Z_q = V^N$. Here $|f_{ij}| \ll 1$ except possibly in a small region about $r_{ij} = 0$. It is therefore sensible to write

$$Z_q = \int (1 + \Sigma f_{ij} + \Sigma\Sigma f_{ij} f_{kl} + \ldots) dq.$$

As a first approximation we reject all terms not linear in the f_{ij}, so that $Z_q = V^N + \int \sum_{i<j} f_{ij}\, dq$. Physically this step corresponds to the notion that when the gas is sufficiently dilute the probability of finding more than one molecule in close proximity to any selected molecule is so small as to be negligible.

Now $J \equiv \int f_{ij}\, dq = V^{N-2} \int f_{ij}\, d\mathbf{r}_i\, d\mathbf{r}_j$. To integrate with respect to $\mathbf{r}_j$, take the origin of coordinates at the ith molecule. Then, writing r for r_{ij} and using polar coordinates, $J = 4\pi V^{N-2} \int d\mathbf{r}_i \int_0^\infty (e^{-\beta u(r)} - 1) r^2\, dr$, having ignored the fact that, strictly speaking, the integration goes only over the interior of the box. This is justified because the only significant contribution to the integral corresponds to very small values of r (i.e. $r \ll V^{\frac{1}{3}}$), and the fraction of molecules lying close to the walls of the box is negligible when V is sufficiently large. Integrating with respect to $\mathbf{r}_i$ gives another factor V. By inspection we now have $Z_q = V^N + \frac{1}{2}N(N-1)J \sim V^N + \frac{1}{2}N^2 J$, so that finally

$$Z_q = V^N (1 - 2\pi N^2 w/V), \qquad w = \int_0^\infty (1 - e^{-\beta u(r)}) r^2\, dr.$$

The equation of state follows as usual from $P = kT\, \partial \ln Z_q/\partial V$:

$$PV/NkT = 1 + 2\pi N w/V.$$

On the right we have the beginning of the virial series. Had we included the effects of the terms $\Sigma f_{ij} f_{kl}$ in our calculations we would have a further term, this time proportional to V^{-2}; and so on. At any rate, since V stood earlier for the molar volume the second virial coefficient is $B_2(T) = 2\pi L w$.

As a schematic example at least, let the molecules be taken as rigid spheres of radius r_o, the forces between them being attractive and so weak that $-\beta u \ll 1$ for $r > 2r_0$. Specifically suppose that $u = -\varepsilon(2r_0/r)^6$ when $r > 2r_0$. (Bear in mind that the distance of closest approach is $2r_0$, not r_0.) Approximating $1 - e^{-x}$ by x we find at once that

$$B_2(T) = 4V_0(1 - \varepsilon/kT), \qquad (V_0 = 4\pi r_0^3 L/3).$$

Remarkably, this form of B_2 coincides exactly with that of a van der Waals gas with $a = 4LV_0\varepsilon$, $b = 4V_0$. Nevertheless this model is far too crude to provide us with a reliable experimental test of the theory. To do any better would, however, lead us into altogether too much detail.

The presence of internal degrees of freedom reveals itself most directly through corresponding contributions to the specific heat C. To calculate them we must evaluate Z_i. Since we are operating under conditions such that the states of the various molecules are independent of each other, $Z_i = z^N$, where $z = \Sigma\, e^{-\beta\varepsilon_j}$, taken over all the internal states of a molecule. Here we have to take into account rotational, vibrational, electronic and nuclear states. As regards the first three, we suppose that H_i is the sum of three corresponding terms H_r, H_v, H_e. (In certain circumstances this approximation is not sufficient, for instance when both rotation and vibration are highly excited.) z now splits up into three separate factors: $z = z_r z_v z_e$; and each factor gives a corresponding contribution to the specific heat. For example, $C_r = Nk\, \partial(T^2 \partial \ln z_r/\partial T)/\partial T$ is the rotational specific heat, and so on. The nuclear states, finally, are accounted for by appropriate weight factors in z_r.

It will suffice to concentrate our attention on diatomic molecules. We first evaluate the sum $z_r = \Sigma\, e^{-\beta\varepsilon_j}$ taken over all the rotational (and nuclear) states. Any particular rotational state is characterized by two integers: j $(= 0, 1, 2, \ldots)$ and m $(= -j, -j+1, \ldots, j)$. Its weight is obviously $2j + 1$, its energy being $\varepsilon_j = j(j+1)h^2/8\pi^2 I = j(j+1)kT_r$, say, where I is the molecular moment of inertia [*Note* 65]. In the simpler case of heteronuclear molecules the nuclear states merely contribute a weight factor $\omega = (2s_a + 1)(2s_b + 1)$ to z_r, s_a and s_b being the nuclear spins of the constituent nuclei. Therefore $z_r = \omega\zeta$, where $\zeta = \sum_{j=0}^{\infty} (2j+1)\, e^{-j(j+1)\tau}$, with $\tau = T_r/T$, and then $C_r = \tau^2\partial^2 \ln \zeta/\partial\tau^2$ in units of Nk. Explicitly,

$$C_r = \begin{cases} 1 + \tau^2/45 + 16\,\tau^3/945 + \ldots & \text{when } \tau \to 0, \\ 12\,\tau^2\, e^{-2\tau}(1 - 6\,e^{-2\tau} + 42\,e^{-4\tau} + \ldots) & \text{when } \tau \to \infty. \end{cases}$$

Evidently $C_r \to 1$ as $T \to \infty$ in agreement with the classical Theorem of Equipartition, whilst $C_r \to 0$ as $T \to 0$ in harmony with the Third Law. Curiously $C_r > 1$ for values of T greater than about $\frac{3}{5}T_r$.

In the case of homonuclear molecules the situation is different since the total wave function of a molecule must be anti-symmetric for fermion nuclei and symmetric for boson nuclei. If their spins are s there are $(2s + 1)^2$ nuclear wave functions of which $(s + 1)(2s + 1)$ are symmetric and $s(2s + 1)$ are anti-symmetric [*Note* 66]. In the case of fermions the symmetric nuclear wave functions must be combined with the anti-symmetric rotational wave functions and vice versa, whereas in the case of bosons the two kinds of nuclear wave functions must be interchanged in this prescription. Therefore, if ζ_+ and ζ_- stand for the sums of those terms of ζ which have even and odd j respectively, whilst z_{r+} and z_{r-} refer to bosons and fermions respectively,

$$z_{r\pm} = (s + 1)(2s + 1)\zeta_\pm + s(2s + 1)\zeta_\mp .$$

For hydrogen H_2 and deuterium D_2 we have z_{r-} with $s = \frac{1}{2}$ and z_{r+} with $s = 1$ respectively, i.e. $z_r(H_2) = 3\zeta_- + \zeta_+$, $z_r(D_2) = 6\zeta_+ + 3\zeta_-$. As a matter of fact, the theory is realistic only for molecules made up of hydrogen, deuterium and tritium because in all other cases T_r is so small that condensation has set in long before a temperature is reached at which the rotational specific heat is no longer describable classically. (For example, $T_r \approx 2.0$ °K for O_2, whereas $T_r \approx 85.4$ °K for H_2.) However, even in the remaining cases of interest there is another complication which arises from the fact that specific heat measurements will normally take place under non-equilibrium conditions on account of the extreme rarity of transitions between different nuclear spin states. In effect, instead of $C_{r\pm} = \tau^2\partial^2 \ln \{[(s+1)\zeta_\pm + s\zeta_\mp]/(2s+1)]\}/\partial\tau^2$ we now have $C_{r\pm} = \tau^2\partial^2\{[(s+1)\ln\zeta_\pm + s\ln\zeta_\mp]/(2s+1)\}/\partial T^2$ [*Note* 67]. Agreement of this last result with experiment is excellent.

We pass on to the vibrational partition function of diatomic molecules. As a first approximation we suppose that the relative motion of the nuclei is simple harmonic with frequency ν, a characteristic temperature $T_v = h\nu/2k$ being then defined. Since $\varepsilon_i = (i + \frac{1}{2})h\nu$ $(i = 0, 1, \ldots)$, $z_v = \Sigma \exp[-(2i + 1)\tau]$, where τ now stands for T_v/T. Provided T is not large compared with T_v the series converges rapidly, so that no harm is done in summing it as it stands even though the terms with very large i are physically unrealistic. Thus $z_v = 1/(2 \sinh \tau)$, and therefore $C_v = \tau^2\partial^2 \ln z_v/\partial\tau^2 = (\tau/\sinh \tau)^2$. This varies from 0 at $T = 0$ to 1 when $T \gg T_v$, as it must. For the common gases the value of T_v, as we have defined it, is typically of the order of one or two thousand degrees K. Notice that when $T \gg T_v$ the required refinements of detail are not at all simple. Not only will the vibration be anharmonic, but we can in any event no longer treat it independently of the rotation, as we already know.

The electronic specific heat comes into play when kT becomes comparable with the amount ε by which the energy of the first excited electronic state (weight ω_1) of the molecule exceeds that of the ground state (weight ω_0). If this time we write $T_e = \varepsilon/k$ and $\tau = T_e/T$, we have $z_e = \omega_0 + \omega_1 e^{-\tau}$, from which we find in the usual way that $C_e = \rho\tau^2 e^{\tau}/(1 + \rho e^{\tau})^2$, with $\rho = \omega_0/\omega_1$. This vanishes for both small and large values of T. On the other hand, taking $\rho = 1$ for example, the maximum value of C_e is about 0.45.

At this point we break off our discussion of gaseous specific heats, for the consideration of various refinements, especially in the context of polyatomic molecules, would not add greatly to our basic knowledge.

Problems

18.1 Let γ stand for the ratio C^*/C of the specific heats at constant pressure and volume respectively. At ordinary pressure and temperature a few measured values are, nearly enough, as follows: Argon, $\frac{5}{3}$; Oxygen, $\frac{7}{5}$; Ethylene, $\frac{5}{4}$. How do you account for these?

18.2 A molecule is made up of s atoms. What is the value of the ratio γ of the specific heats, calculated classically?

18.3 Around 180 °K the molecule of nitric oxide NO is effectively rigid. At 178 °K the measured value of C is about 2.68. Now, the NO molecule has an excited electronic state whose energy exceeds that of the ground state by an amount 178 k, both states having weight 2. Calculate C at 178 °K and compare your result with the measured value.

18.4 At what temperature are the equilibrium abundances of ortho- and para-hydrogen equal?

18.5 Atomic nuclei can themselves exist in various excited states, yet we have not considered any corresponding contribution to the internal partition function. Why not?

LECTURE 19

Ideal Fermion and Boson Gases

In the course of the previous lecture we saw how the introduction of quantum mechanical ideas into statistical thermodynamics overcame the discrepancy between predicted values of specific heats and those that are observed. On the other hand, the conflict between the Third Law and the conclusion that systems of non-interacting particles have the equation of state $PV/T =$ constant remained. We naturally want to see in some detail how this is resolved by describing also the translational motion quantum mechanically, not classically as we did last time. To have everything as simple as possible we therefore study a system of non-interacting non-relativistic particles in a rigid box of volume V, the particles being taken as structureless except in as far as they may have intrinsic spin.

The easiest way to proceed is through the grand canonical partition function Z. From Lecture 17 we know that $Z = \sum_N \lambda^N \sum_j \exp(-\beta E_{Nj})$, where $E_{Nj} = \sum_i n_{ji}\, \varepsilon_i$ and $\sum_i n_{ji} = N$. It is most convenient to write $\eta_i = \exp[\beta(\mu - \varepsilon_i)]$, for then $Z = \sum_N \sum_j \prod_i \eta_i^{n_{ji}}$. To sum over j means in effect to sum over those sets of values of the occupation numbers which are consistent with the condition $\sum_i n_{ji} = N$. However, since we must subsequently sum over N anyway, the formation of the double sum in Z is equivalent to summing over the occupation numbers independently of each other, so that $Z = \prod_i \sum_{n_i} \eta_i^{n_i}$. We now have two alternative possibilities, namely, either the particles are fermions in which case every $n_i = 0$ or 1, or else the particles are bosons and then the n_i are unrestricted. At once $Z = \Pi(1 \pm \eta_i)^{\pm 1}$, the alternative signs going together, $+1$ and -1 referring to fermions and bosons respectively [*Note* 68]. The grand potential is therefore

$$X = \pm kT \sum_i \ln(1 \pm \eta_i).$$

During our investigation of the classical limit we found that $Z = e^{\lambda z} = \prod_i e^{\eta_i}$. Comparison with the general result therefore shows that in the classical limit we must have $\eta_i \ll 1$, that is to say, the condition $\lambda \ll 1$ must be satisfied. In short [*Note* 69], the condition that quantum effects can be disregarded is that $\zeta \equiv \bar{N}l^3/\omega V \ll 1$. When this is the case we call the gas *non-degenerate*.

Now, from the general relation $\bar{N} = \partial X/\partial\mu$ we get $\bar{N} = \Sigma\, \eta_i/(1 \pm \eta_i)$. Since $\bar{N} = \Sigma\, \bar{n}_i$, this suggests that the mean occupation number of the level with energy ε_k is $\bar{n}_k = \eta_k/(1 \pm \eta_k)$.

In fact, we can obtain directly this result and the expression $(\delta n_k/\bar{n}_k)^2 = 1/\eta_k = 1/\bar{n}_k \mp 1$ for the relative mean square fluctuation of n_k directly by differentiating Z with respect to η_k [*Note* 70]. In particular, the relative fluctuation of the occupation number of a highly occupied boson level is of the order of unity.

From this point onwards we treat fermion and boson systems separately because, as we shall see, they exhibit quite different behaviour when they are degenerate. For fermions we have $X = kT\,\Sigma \ln(1 + \lambda\, e^{-\beta\varepsilon_i})$. Since $\bar{N}$ is large and each state can be occupied at most once, the majority of particles will occupy states whose energy is much greater than that of the ground state. In replacing summation by integration the error implicit in the fact that the smoothed-out density of states $g(\varepsilon)$ vanishes for the ground state can therefore be ignored. Thus

$$X = 2\pi\omega(2mh^{-2})^{\frac{3}{2}}(kT)^{\frac{3}{2}}V\int_0^\infty t^{\frac{1}{2}} \ln(1 + \lambda\, e^{-t})dt,$$

having set $\varepsilon = kTt$. Integrating by parts and introducing the thermal wavelength l, this becomes

$$X = \omega l^{-3}\, kTV\,\xi(\lambda), \qquad \xi(\lambda) = \tfrac{4}{3}\pi^{-\frac{1}{2}}\int_0^\infty [t^{\frac{3}{2}}/(\lambda^{-1}\, e^t + 1)]dt.$$

Almost classical behaviour corresponds to sufficiently small values of λ. In that case we can write $1/(\lambda^{-1}\, e^t + 1) = \lambda\, e^{-t} - \lambda^2\, e^{-2t} + \ldots$ and integrate term by term:

$$\xi(\lambda) = \lambda - 2^{-\frac{5}{2}}\lambda^2 + 3^{-\frac{5}{2}}\lambda^3 + \ldots.$$

$X(V, T, \mu)$ is now known explicitly and we can get the equation of state from it in the usual way be eliminating λ between $\partial X/\partial\mu = \bar{N}$ and $P = X/V$. It takes the form of a virial expansion in powers of ζ, the leading terms being $PV/\bar{N}kT = 1 + (1/4\sqrt{2})\zeta + \ldots$.

The case of the highly degenerate gas is more interesting. This time λ is very large. Observe now that as $\lambda \to \infty$ the factor $1/(\lambda^{-1}\, e^t + 1)$ is very nearly a step function whose value goes from 1 to 0 as t increases through the value $\ln\lambda$. In this limit therefore $\xi(\lambda) \sim (4/3\sqrt{\pi})\int_0^{\ln\lambda} t^{\frac{3}{2}}\, dt = (8/15\sqrt{\pi})(\ln\lambda)^{\frac{5}{2}}$. A better approximation is [*Note* 71] $\xi(\lambda) \sim (8/15\sqrt{\pi})(\ln\lambda)^{\frac{5}{2}}[1 + (5\pi^2/8)(\ln\lambda)^{-2} + \ldots]$. Therefore

$$X = (8\pi\omega/15)(2mh^{-2})^{\frac{3}{2}}\,\mu^{\frac{5}{2}}\,V[1 + (5\pi^2/8)(kT/\mu)^2 + \ldots].$$

In the limit $T \to 0$ we get from this $\bar{N} = \partial X/\partial\mu = (4\pi\omega V/3)(2mh^{-2}\,\mu)^{\frac{3}{2}}$. The value of μ defined by this equation is the *Fermi energy* ε_F, which in turn defines a temperature $T_F = 2\varepsilon_F/5k$. Going on to the next approximation, the equation of state now takes the form

$$PV = \bar{N}kT_F\,[1 + (\pi^2/15)(T/T_F)^2 + \ldots],$$

whilst the specific heat $C = (\pi^2\,\bar{N}k/5T_F)\,T + \ldots$ vanishes linearly with T as $T \to 0$, as does the entropy. The demands of the Third Law are thus fully met here. Of course, we must not forget that T_F itself depends upon V. When $T = 0$ the equation of state of an electron gas ($\omega = 2$), for instance, is, with $\mathrm{v} = V/\bar{N}$, $P\mathrm{v}^{\frac{5}{3}} = (3/\pi)^{\frac{2}{3}}\,h^2/20\,m$, from which Boltzmann's constant is characteristically absent.

If, on a qualitative level, we consider the conduction electrons in metals to constitute a free electron gas, it turns out that T_F is commonly comparable with 10^5 °K, so that the gas will be completely degenerate at ordinary temperatures and its contributions to the specific heat of the metal is negligible. At, say, 1000 °K the specific heats of many metals indeed closely approximate the value $3\bar{N}k$ predicted by the classical theorem of equipartition. Were the electron gas not degenerate this predicted value would be $3(1 + \frac{1}{2}\nu)\bar{N}k$, where ν is the number of conduction electrons per atom. Gas degeneracy also plays a crucial role in the description of the internal structure of certain stars to which we shall return very briefly later on.

It remains to consider boson systems, concentrating our attention on the low temperature case. At first sight we might be tempted to retain our previous expression

$$X = \omega l^{-3}\, kTV\xi(\lambda)$$

provided we now take $\xi(\lambda) = \frac{4}{3}\pi^{-\frac{1}{2}} \int_0^\infty [t^{\frac{3}{2}}/(\lambda^{-1}\, e^t - 1)]dt$. However, we have already noted that as $T \to 0$ all particles will now end up in the ground state, so that the assignment to this of the weight zero by the smoothed-out density of states $g(\varepsilon)$ is quite unacceptable. What we must therefore do is to retain the first term of the series $-kT\,\Sigma \ln(1 - \eta_i)$ explicitly and replace summation by integration only for the remaining terms. As a result we now have

$$X = -\omega kT \ln(1 - \eta_0) + \omega l^{-3}\, kTV\, \xi(\lambda).$$

The integral $\xi(\lambda)$ will obviously converge only if $\lambda \leqslant 1$, i.e. $\mu \leqslant 0$. Under these circumstances $\xi(\lambda) = \frac{4}{3}\pi^{-\frac{1}{2}} \int_0^\infty t^{\frac{3}{2}}(\lambda e^{-t} + \lambda^2 e^{-2t} + \ldots)dt = \sum_{s=1}^{\infty} s^{-\frac{5}{2}}\lambda^s$. Since $\varepsilon_0 \to 0$ as $V \to \infty$ we can set $\varepsilon_0 = 0$. Therefore now $\bar{N} = \partial X/\partial\mu = \omega\lambda/(1 - \lambda) + \omega l^{-3}\, V \sum_{s=1}^{\infty} s^{-\frac{3}{2}}\lambda^s$.

Next, write $\theta(\lambda)$ for the ratio of the sum on the right to its value, a say, at $\lambda = 1$, and define a temperature T_B by the relation $N' = (2\pi mh^{-2}\, kT_B)^{\frac{3}{2}}Va$, with $N' = \bar{N}/\omega$. The equation for λ then takes the simpler form $\lambda/(1 - \lambda) = N'[1 - \theta(\lambda)(T/T_B)^{\frac{3}{2}}]$. When $T < T_B$ and $\bar{N}$ is sufficiently large we can solve this by iteration. Writing $T/T_B = \tau$, we get $\lambda = 1 - 1/[N'(1 - \tau^{\frac{3}{2}})] + O(N'^{-2})$. Since $\bar{n}_0 = \eta_0/(1 - \eta_0) = 1/(\lambda^{-1} - 1)$ it follows from this that $\bar{n}_0/N' = (1 - \tau^{\frac{3}{2}}) + O(1/N')$. Taking the limit $N' \to \infty$, $\lambda = 1$ and $\bar{n}_0/N' = (1 - \tau^{\frac{3}{2}})$. On the other hand, in this limit the equation for λ shows that when $T > T_B$ the factor in square brackets must vanish, which means that λ is now determined by the equation $\theta(\lambda) = \tau^{-\frac{3}{2}}$. This may be solved in series: $\lambda = a\,\tau^{-\frac{3}{2}} - (a^2/2\sqrt{2})\,\tau^{-3} + \ldots$.

The gas pressure is X/V. On inserting the results just obtained, we find at once that when $T < T_B$ the term $-(kT/V)\ln(1 - \eta_0)$ vanishes in the thermodynamic limit. Therefore $P = \kappa\bar{N}k\, T_B^{-\frac{3}{2}}T^{\frac{5}{2}}V^{-1}$, where $\kappa = \xi(1)/a \approx 0.5134$. (Note that P is in fact quite independent of $V/\bar{N}$.) Further, $U = \frac{3}{2}PV$, so that the specific heat, in units of $\bar{N}k$, is $C = \frac{15}{4}\kappa\tau^{\frac{3}{2}}$. When $T = T_B$ this exceeds its classical value by an amount $\frac{3}{2}(\frac{5}{2}\kappa - 1) \approx 0.425$. When $T > T_B$ the differentiations are more cumbersome because λ now depends upon T. At any rate, it turns out that the various quantities just considered are all continuous at $T = T_B$, and that C decreases monotonically to its classical value as T increases beyond T_B. However, dC/dT has a finite discontinuity at $T = T_B$, so that the ideal boson gas undergoes a third order transition at that temperature. As we have seen, below it a large fraction of the particles begins to accumulate in the ground state, and then no longer contributes to the energy, pressure, or entropy. This process of accumulation is known as the *Einstein condensation*, for below T_B one effectively has a mixture of two phases. Liquid helium consisting of the normal isotope He^4 (which is a boson) has $T_B \approx 3.1$ °K. Remarkably it undergoes a transition at about 2.2 °K (at which a superfluid component begins to appear), but unfortunately the transition appears to be of the second order. In any event, liquid helium is far from being an ideal gas, and in the light of theoretical uncertainties and empirical discrepancies the true nature of the transition remains obscure.

Finally a few words about black body radiation which is, by definition, a radiation field in diathermic equilibrium with its surroundings. We may think of it as a photon gas, a photon being regarded as a 'particle' of zero rest mass. Since the exclusion principle does not operate we are dealing with bosons, and, since a photon has two possible independent directions of polarization, $\omega = 2$. Now, the energy ε of a photon of momentum $\mathbf{p}$ is $c|\mathbf{p}|$, where c is the speed of light. Therefore we can obtain the appropriate density of states through the formal substitution $\varepsilon = |\mathbf{p}|^2/2m \to \varepsilon^2/2mc^2$ in our previous expression for $g(\varepsilon)d\varepsilon$. As a result we must have $g(\varepsilon)d\varepsilon = 8\pi(ch)^{-3}\,V\varepsilon^2\,d\varepsilon$, or equivalently $g(\nu)d\nu = 8\pi c^{-3}\,V\nu^2\,d\nu$, since $\varepsilon = h\nu$ for a photon of frequency ν. The Gibbs function of black body radiation vanishes identically, so that $\mu = 0$ [*Note* 72]. Therefore we now have $X = -kT\,\Sigma \ln(1 - e^{-\beta\varepsilon_i})$, but here nothing prevents us from replacing summation by integration since there is now no fixed total number of particles and the previous complication of the 'condensation' into the ground state does not arise. Thus

$$X = -8\pi kTc^{-3}\,V\int_0^\infty \nu^2 \ln(1 - e^{-\beta h\nu})d\nu,$$

which becomes, on integrating by parts, $X = PV = \frac{1}{3}U = (8\pi/3c^3)hV\int \nu^3 d\nu/(e^{-\beta h\nu} - 1)$. Since the mean occupation number $\bar{n}_\nu$ is given by $1 + 1/\bar{n}_\nu = 1/\eta_\nu = e^{h\nu/kT}$, we can decompose the last result as follows: the energy density $u(\nu)d\nu$ of black body radiation in the interval $(\nu, \nu + d\nu)$ is given by

$$u(\nu) = 8\pi c^{-3}h\nu^3/(e^{h\nu/kT} - 1).$$

which is just *Planck's radiation law*.

Problems

19.1 Express the constant σ which occurs in Note 72 in terms of the constants c, h and k.

19.2 Explain the remark made during the lecture to the effect that one would not expect Boltzmann's constant k to appear in the equation of state of a fermion gas at $T = 0$.

19.3 Derive the equation of state of a fermion gas at $T = 0$ without using statistical methods as such.

19.4 Find an expression for the energy ε_F of the highest occupied level of a completely degenerate relativistic fermion gas. In particular consider the limits $N/V \to 0$ and $N/V \to \infty$.

19.5 Obtain an equation which relates the Fermi temperature T_F to the temperature T_0 at which the chemical potential of a fermion gas is zero, the density of states being $g(\varepsilon)$. In particular, show that the ratio of the two temperatures is constant when $g(\varepsilon) = A\varepsilon^r$, where A and r are positive constants.

LECTURE 20

Historical Remarks. Negative Absolute Temperatures

No lecture course is complete without an historical reminder, and this is a good point at which to issue it. Last time we learned that the density of states in an isotropic radiation field is $g(\nu) = 8\pi c^{-3} V\nu^2$. What we are counting is the number of standing electromagnetic waves with frequencies in the range $(\nu, \nu + d\nu)$ in a perfectly reflecting enclosure. Each such wave is formally a linear harmonic oscillator and in equilibrium the Theorem of Equipartition of Energy assigns an energy kT to it. The total energy of the field is therefore $U = kT \int_0^\infty g(\nu)d\nu = \infty$, a conclusion which is entirely unacceptable. The classical laws clearly need to be modified, and it was this recognition which gave rise to the beginnings of quantum theory. In a sense we have here followed an analogous route, but of course things were much easier for us, to the extent that we were able to fall back on quantum mechanics as a fully-fledged theory. Still, it is not surprising that we were able to resolve the conflicts which arose, such as that between the Third Law and the ideal gas equation $PV = RT$.

Interestingly enough, historically the first application of the theory of degenerate fermion gases was in an astrophysical context. There is a class of small, faint, very dense stars—white dwarf stars—whose observed characteristics originally could not be explained. Now, these stars consist largely of fully ionized helium at temperatures of the order of 10^7 °K, the electron density being of the order of 10^{30} cm^{-3}. The Fermi temperature of the electrons is then of the order of 10^{10} °K which far exceeds the temperature of the stellar material. Accordingly the electron gas is virtually completely degenerate. Its pressure is therefore so great that that of the helium nuclei can be disregarded. The theory which takes these features into account is entirely successful, but we have not the time to develop it here.

It is only fitting that to end our course we should return briefly to the phenomenological theory—without hesitating to draw upon the conceptions of the statistical theory for further enlightenment. Our aim is to re-examine certain interrelated assumptions which we made earlier. They were (i) that all reversible transitions must be quasi-static, which led us, indeed, to treat the terms 'quasi-static' and 'reversible' as synonymous; (ii) that any state of a given system K (not adiabatically isolated) could always be reached quasi-statically from any other, with the conclusion that T cannot be negative; (iii) that the energy of K could be increased indefinitely by some weakly isometric process. What if for special systems one or all of these should happen to be false?

In the first place, the generic framework of our theory would not be affected. In particular, recall that the existence of the empirical entropy function merely subsumes the relative adiabatic accessibility of pairs of states; but there is no implication that mutual accessibility must be through continuous sequences of states. In short, we must now sharply distinguish between reversible and quasi-static transitions and likewise between non-static and irreversible transitions. Moreover, if we have an 'abnormal system'—let us call it K^*—for which the second assumption is false then its states will divide themselves into two classes in such a way that two states will be quasi-statically accessible from each other only if they belong to the same class. The argument we used in the ninth lecture to show that the sign of T is fixed then covers only each class separately, so that if T is positive for one class there is nothing to say that it cannot be negative for the other. That the ideal gas scale 'comes to an end' at $T = 0$ is entirely irrelevant to such a situation. This scale does not *define* the absolute temperature, it merely happens to coincide with it when $T > 0$: the function $T(t)$ enters the phenomenological theory as an integrating denominator of dQ and in no other way. Equally irrelevant is the related objection that the varying degrees of molecular motion just exhaust the positive values of T, leaving no room for negative values. Indeed, from this we can infer no more than that systems with inherent 'molecular motions' can only have $T > 0$.

Having now reverted to the statistical picture, we recall that generically T does not belong to a given system but rather to its representative ensemble taken as a whole. There is no *a priori* reason why it should have one sign rather than another. In principle its sign is only limited by the need for the sum $Z = \Sigma\, e^{-E_j/kT}$ to converge. It follows that T must be positive *unless* the sequence E_j breaks off after a finite number of terms. When this is the case U is finite for all values of T so that the third of the assumptions under review is then also not justified.

The question is now whether abnormal systems K^* are physically realizable in any reasonable sense of the term. The number of states must now be finite. This immediately brings to mind a particle with given intrinsic spin and magnetic dipole moment. When such a spin j particle is situated in a fixed external homogeneous magnetic field $\mathbf{B}$ its magnetic moment μ can assume only $2j + 1$ possible orientations relative to the direction of $\mathbf{B}$. Its corresponding energy levels are $-\gamma mB$, where γ is a constant, $B = |\mathbf{B}|$, and m varies by integral steps from $-j$ to j. Each atomic nucleus, situated at a vertex of a crystal lattice is just such a particle. All the nuclei of a given kind constitute a 'nuclear spin system' which has a finite number of (magnetic) energy levels in the field $\mathbf{B}$. These are $-N_m m\gamma B$, where N_m is the number of spins with orientation specified by m. Here we have supposed the interactions between the various spins to be so weak that their contribution to the energy can be disregarded. Nevertheless they exist, and because of this the numbers N_m will change until in a time of the order of θ_m, say, equilibrium has established itself. Then we can describe the situation by the usual statistical method. Thus $Z = z^N$, where $z = \sum_{m=-j}^{j} \exp(m\gamma B/kT) = (\sinh g\xi)/\sinh \xi$, where $\xi = \gamma B/2kT$ and $g = 2j + 1$. By differentiation of $F = -NkT \ln z$ it follows that

$$S = Nk\{\ln[(\sinh g\xi)/\sinh \xi] - \xi(g \coth g\xi - \coth \xi)\},$$

$$U = -\tfrac{1}{2}N\gamma B(g \coth g\xi - \coth \xi), \qquad C = Nk\xi^2(\operatorname{cosech}^2 \xi - g^2 \operatorname{cosech}^2 g\xi).$$

At this point we must digress for a moment. When the crystal is at temperature T_l the nuclei will carry out their usual thermal motions under the influence of mutual elastic (electronic) restoring forces. These motions are formally collective vibrations of the crystal as a whole, the so-called lattice vibrations. Owing to the loose coupling which exists between the nuclear spin system K^* and the lattice vibrations the temperature which enters into the thermodynamic quantities of K^* just calculated will be T_l, granted of course that the crystal as a whole is in true, overall equilibrium. This reservation we shall shortly re-examine. At any rate, as T ranges from 0 to ∞ the entropy S of K^* goes from 0 to $Nk \ln g$. Its energy U goes from $-Nj\gamma B\,(= -U_0$, say) to 0, the first value corresponding to all the spins being aligned in the direction of **B**, the second to all spin directions being equally represented.

So far we seem to have encountered nothing very unusual except that U has the upper bound zero. This means in particular that no amount of 'stirring'—say by exposing K^* to appropriate radio-frequency electromagnetic pulses—can increase the energy beyond this limit, even though T may tend to infinity. Now recall that we took T to be positive because we equated T with the necessarily positive temperature T_l, having taken the crystal to be in overall equilibrium. Compare this situation with the following: a gas contained in an enclosure made of some substance such as polystyrene. It is a matter of experience that overall equilibrium will establish itself only very slowly. Then we consider ourselves entitled to regard the gas by itself as 'the system', to which a definite temperature can be assigned. The surrounding material constitutes an 'adiabatic enclosure', and to this also a definite temperature may be assigned which need not be the same as that of the gas. This is of course the usual idealization, but it reflects practical realities. What is crucial is that the establishment of equilibrium between 'system' and 'enclosure' should take place only during a time interval very much longer than the time it might take to subject the gas to any effectively quasi-static transition of interest. In other words, conceptually the situation makes sense only if the thermal coupling between systems and surroundings is sufficiently weak.

By analogy we may look upon the lattice vibrations K_l of a crystal to be the 'surroundings' of the 'system' K^* of nuclear magnetic moments. Provided the coupling between K_l and K^* is so weak that the establishment of equilibrium between them takes a time of the order of θ_l, say, with $\theta_l \gg \theta_m$, K^* will in effect be adiabatically isolated. During a time θ such that $\theta_m \ll \theta \ll \theta_l$ a temperature T can be assigned to K^*, but this need no longer be equal to T_l [*Note* 73].

Now, after a crystal has been situated in the field **B** for a time much longer than θ_l—so that $T = T_l$—let the direction of the field suddenly be reversed [*Note* 74]. The effect of this reversal is the same as that of the notional reversal of all the spins. The energy of K^* has therefore reversed its sign, and since equilibrium still obtains [*Note* 75] the temperature of K^* is now negative. In short, K^* has undergone a non-static transition from a state $\mathfrak{S}'$ $(T' > 0)$ to a state $\mathfrak{S}''$ for which $T'' = -T'$, $U'' = -U'$ and $S'' = S'$. Consistently with the last equation this transition is nevertheless reversible—we need only suddenly reverse the field again. Quasi-statically the states $\mathfrak{S}'$ and $\mathfrak{S}''$ are, however, mutually inaccessible, bearing in mind our previous conclusion that T can never reverse its sign in a quasi-static transition. Evidently when $T < 0$ the energy of K^* is greater than its energy at any positive temperature. Moreover, heat will pass *from* a system with $T < 0$ *to* a system with $T > 0$ when these are placed in diathermic contact. It is therefore entirely natural to adopt a standard empirical temperature scale $t = -1/T$, for then, once and for all, that of two systems is the hotter which has the higher (standard) empirical temperature.

At the same time, both S and U will depend continuously upon t, U decreasing monotonically from its maximum U_0 to its minimum $-U_0$ as t ranges from $-\infty$ to ∞, whilst S is an even function of t with its maximum $Nk \ln g$ at $t = 0$. Finally, the limit $T \to 0$ with which the Third Law concerns itself splits up into the two limits $t \to -\infty$ and $t \to +\infty$: in both S must tend to a finite limit which is independent of the d-coordinates.

Our discussion of negative absolute temperature states—somewhat ephemeral though they may be in practice—has been well worth while. If nothing else it has once and for all laid to rest the notion that conceptually the absolute temperature is inevitably bound up with the ideal gas scale: it is not. To believe otherwise is merely to cling to a prejudice which only gives rise to further prejudices; and that seems a suitable remark with which to end this course.

Problems

20.1 Write down S, U and C for K^* as functions of T when $j = \frac{1}{2}$.

20.2 Take $j = \frac{1}{2}$. Find the extreme of the specific heat of K^*.

20.3 Is it conceivable that a system consisting of a very large number of particles could not be describable by the methods of statistical mechanics, in the sense that the temperature of such a system could be neither positive nor negative, i.e. T would not be meaningful?

20.4 At the end of the seventh lecture we noted the impossibility of a 'perpetual motion machine of the second kind'. Is it possible to have such a negative temperature machine?

20.5 Paramagnetic salts are often cooled by demagnetizing them adiabatically. Can a nuclear spin system be demagnetized in this way when (i) $T > 0$, (ii) $T < 0$?

NOTES

Note 1

This is an excellent example of a situation in which one has to be 'reasonable', i.e. pragmatic. Taken literally the definition is empty, for *every* system changes in the course of time for one reason or another. For instance, a terrestrial laboratory is affected by the tides and, in any event, will be disastrously affected by future changes in the sun. We may be prepared to ignore these factors, the one on the grounds of the smallness of the effects in question, the other on the grounds that it is irrelevant to the present time. Nevertheless the difficulty persists. A block of ordinary glass will usually show no measurable changes over a time span of many years, yet we know that it is capable of changing spontaneously from an amorphous to a crystalline form: so that initially it cannot have been in equilibrium. (At best, the glass could be treated as if it were in equilibrium only in experiments which do not affect its generic structure.) In short, we have no option but to assume that a system which has every *appearance* of being in equilibrium *is* in fact in equilibrium. When a clash between theory and experiment arises we must then enquire whether just the assumption that equilibrium obtained initially was perhaps erroneous.

Note 2

Superficially this deformation seems to be different from the kind of deformation encountered earlier since we might argue that we are deforming the container, and that it is not part of the system. We can save the situation by noting that at the same time different parts of the gas are displaced relatively to each other.

Note 3

Such a statement may look rather bewildering at first sight, considering that elsewhere one does apparently meaningfully discuss 'non-equilibrium states'. There is no contradiction. In one case we are operating strictly within the framework of the definitions of one theory, in the other one is concerned with the definitions of another theory, e.g. fluid dynamics, and then what is meant by a 'state' is something altogether different.

Note 4

The phrase 'at an infinitesimal rate' of course cannot be taken literally since otherwise we would be confronted with the unsavoury conclusion that a quasi-static transition occupies an infinite time interval. As in *Note* 1 we face the need to adopt a sensible interpretation. Here a transition proceeds 'at an infinitesimal rate' if it proceeds so slowly that the deviations from equilibrium are such that the system nevertheless behaves as if it were in equilibrium, i.e. to within experimental error the disequilibrium has no observable consequences.

Note 5

Physicists often rely on intuition for an understanding of this definition. That is, indeed, often quite good enough. What is in fact usually meant when speaking of 'the neighbourhood' is 'a neighbourhood of' or more precisely 'an ε-neighbourhood of', ε being some preassigned positive number. Having chosen a set of coordinates of K and the units in which these are to be measured, it will suffice to take an ε-neighbourhood of $\mathfrak{S}[x_1, \ldots, x_n]$ to consist of all states $\mathfrak{S}'[x'_1, \ldots, x'_n]$ such that $|x'_1 - x_1| + |x'_2 - x_2| + \ldots + |x'_n - x_n| < \varepsilon$. In the context of infinitesimal transitions, or 'transitions between neighbouring states', ε must be taken to be sufficiently small. Thus, if something is said to be true 'in every neighbourhood of $\mathfrak{S}$' it means that it is true no matter how one chooses ε; in particular that it will be true for arbitrarily small values of ε.

Note 6

We are again confronted with an idealization and we cannot be sure that it is warranted. After all, it is certainly true that in practice we cannot construct an 'ideal' thermos flask. We may well think it 'reasonable' to allow its introduction into the theory, especially as only a question of time scale appears to be involved and we already encountered difficulties surrounding this several times before. Still, one has to be careful: what is 'reasonable' today may not be 'reasonable' tomorrow. Sudden changes in attitude have occurred in the history of physics, for example following upon the introduction of the Heisenberg Uncertainty Principle, the Restricted Relativity Principle, and the like.

Note 7

Since the passage of an electric current through K was taken to constitute mechanical interactions, this statement would appear to be true when the source of the current is an electric generator but false when it is a battery. To be on the safe side we should supplement 'motions' with the words 'or chemical reactions'.

Note 8

The first of them is usually known as the Principle of Carathéodory. It is largely equivalent to the traditional statements of the Second Law of Thermodynamics. Its experimental confirmation is mostly indirect: no-one in his senses would suppose that, as a practical proposition, its validity could be tested directly. In any event, such a remark might be made about many laws of physics.

Note 9

Even now it should be emphasized that the function s is so far defined only on the discrete set of states $\mathfrak{S}_1$, $\mathfrak{S}_2$, ... Questions of its continuity or differentiability are therefore as yet beside the point.

The construction we have carried out could of course be undertaken also in the context of other systems for which 'states' have been defined, as, for example, in the case of a dynamical particle system, the state being taken as a set of values of position coordinates and of the corresponding velocities. However, the construction is then empty in the sense that all representative points will coincide, i.e. s is a fixed constant. To a large extent the essential content of the Second Law (to be stated shortly) is that for thermodynamic states the construction is, in fact, not empty.

Note 10

That the set of possible states of K_0 is continuous whereas the set of quantum mechanical states of a bound system is discrete must on no account be thought to involve a contradiction: quantum states are not the states we are talking about, and a given 'quantum state' indeed depends continuously on the values of the d-coordinates which go to make up a thermodynamic state when these vary at an infinitesimal rate.

Note 11

By a 'good' function I mean a function which is single-valued, continuous and sufficiently often differentiable, at least within some finite neighbourhood of a given state of interest. Here again it is best to postulate that s is good. Given only the existence of states inaccessible from a given state and the non-existence of pairs of mutually inaccessible states, together with assumptions about neighbourhoods of states one can, by sophisticated arguments, prove the existence of a continuous function s. However, one still cannot prove its differentiability, so that the whole endeavour is in the end just a wasteful exercise. In any event, functions which turn up in physics are often simply assumed to be good: whether they really are or not can obviously not be tested by direct experiment. Where discontinuities appear to occur one simply has to analyse them by special arguments.

Note 12

This statement is in practice not rigorously true. It represents an idealization in which at least the assumption that all elastic substances in the system are 'perfectly elastic' is satisfied. As remarked earlier, for instance in *Note* 6, one always has to be careful with idealizations; in the present context not least because one runs the risk of a circular argument which in the end says no more than that a certain statement is true when it is true. What it all amounts to here is this: in a strictly mechanical system, all elastic substances must be perfectly elastic, and a perfectly elastic substance is such that the amount of work done on it depends solely upon the deformation it has undergone. This therefore is a definition, and we can then determine by experiment whether a given substance is perfectly elastic or not. In practice it must of course suffice that it should, under appropriate circumstances, have the property only 'to within experimental error'.

Incidentally, the system must be both initially and finally at rest. Evidently any change in configuration must occur 'sufficiently slowly' or else vibrations will be set up and the only way to stop these is by means of frictional forces. Then the statement under discussion would again be false.

Note 13

A conservation law governing some physical quantity q declares that as the system with which it is associated pursues its history the value of q remains constant; or sometimes, more weakly, that the value of q in the remote future will be equal to its value in the remote past. Thus, given the initial value of q one knows its value at all times (or at least in the remote future) without necessarily having to know anything about the details of the evolution of the system, details which may be unavailable in practice or sometimes even in principle.

It should perhaps be mentioned that our 'conservation of energy of mechanical systems' appears to be in conflict with conventional classifications in which one sometimes speaks of 'dissipative (or non-conservative) mechanical systems'. However, the conflict is merely one of semantic detail: the usual 'dissipative mechanical system' is in fact a thermodynamic system. A corresponding difficulty appears to exist with regard to systems with time-dependent constraints. This can also easily be disposed of. In any event, in both cases we have something more general than the systems of interest to us here in that the details of their dynamical histories are involved.

Note 14

Of course it may happen that the adiabatic transition is impossible, as we already know. However, since all that is involved is the value of W_0, we need only mutually interchange the terminal states and then measure the amount of work, i.e. $-W_0$, done in any adiabatic transition between them.

Note 15

Here this says no more than that if, overall, no work is done in an adiabatic transition then U is constant (i.e. $\triangle U = 0$), irrespectively of the details of the transition. The law of conservation of energy therefore essentially amounts, in the present context, to the possibility of defining U (and earlier E) at all.

Note 16

The experimental determination of the function $\tau_A(x)$ contains an inevitable hiatus. The best one can do is to find a necessarily finite number of states with the required property and then construct a function which will accommodate them all. This is essentially a process of interpolation which is at least consistent with the assumption that τ_A is a good function. Nevertheless, although we are looking for a function which is 'known' to exist we can never be sure that we have indeed ended up with the 'right' function, however much we may be influenced by notions of simplicity and the like. These remarks are of course relevant to many other functions of physical theory.

Note 17

Emphatically, temperature is only defined when equilibrium obtains. Familiar expressions such as 'irreversible isothermal transition', for example, are therefore meaningless, unless one interprets 'isothermal' in the weak sense that merely the initial and final temperatures must be the same. As long as we adhere strictly to our definitions, we are even forced to reject Fourier's equation governing the distribution of temperature in a given body. One cannot have 'different temperatures at different points' since this phrase is inconsistent with the condition of equilibrium which is required if the temperature is to have a definite meaning in the first place. (See, however, the answer to Problem **6.4**.)

Note 18

We recall that the increase of P with U is here an experimental fact. It must not be thought that the same behaviour would necessarily obtain whatever the substance within the cylinder might happen to be. If, for example, we were to start with a mixture of ice and water at atmospheric pressure, the effect of increasing U (and t), say by turning the stirrer, would be to decrease the pressure at first, though if we kept stirring it would eventually increase again. Under these circumstances we evidently do not have a good temperature function since it is not single-valued.

Note 19

A symbol such as dQ is to be strictly regarded as a *composite* symbol. Except in special circumstances one cannot look upon it as representing the action of a differential operator d on some function Q. A simple example will make this point quite clear. Given, say $dQ = ydx + zdy + xdz$, there is certainly no function $R(x, y, z)$ of which dQ is the total differential, for if there were we should have to have $\partial R/\partial x = y$, $\partial R/\partial y = z$, $\partial R/\partial z = x$. Differentiating the first and second of these with respect to y and x respectively, one gets an obvious inconsistency. Except in special situations one cannot even find two functions, w and R, say, such that $dQ = wdR$. This is so, for instance, in the case of the example just considered.

Note 20

Thus, whenever the ratios of the differentials are such as to make $dQ = 0$ they make $ds = 0$ at the same time. Assuming that $X_n\, dx_n \neq 0$ without loss of generality, $dQ = 0$ entails that $dx_n = -\sum_{k=1}^{n-1} (X_k/X_n)\, dx_k$. Then $ds = \sum_{k=1}^{n-1} (s_k - s_n X_k/X_n)dx_k$, and this must vanish for all choices of the dx_k $(k \neq n)$. Hence $X_k/s_k = X_n/s_n$ $(k = 1, \ldots, n-1)$ which is already the required result.

Note 21

Evidently every dQ has a generic form which falls under the heading of the 'special situations' referred to at the end of *Note* 19. λ and s are of course not uniquely determined, for if s and λ are one pair such that $dQ = \lambda ds$, then $s^* = f(s)$ and $\lambda^* = \lambda[df(s)/ds]^{-1}$ is another, where f is any (good, monotonic) function whose arbitrariness reflects that of the empirical entropy scale. Incidentally, the factor λ is often called an *integrating denominator* of ds.

Note 22

This means that dQ/T is a total differential (see the preceding *Note*). The use of t as h-coordinate is of course not mandatory. If empirical temperature functions $t_A(x)$, $t_B(y)$, ... (on a common scale) of systems K_A, K_B, ... have been obtained then, for any choice of coordinates, $dQ_A/T(t_A(x))$, $dQ_B/T(t_B(y))$, ... are total differentials of functions $S_A(x)$, $S_B(y)$, ..., respectively.

Note 23

This conclusion is obvious as long as K_A and K_B remain adiabatically isolated from each other since then S_A and S_B separately cannot decrease. It is, however, valid under much wider conditions, namely when transitions or parts of transitions take place as the result of K_C being modified either by the replacement of the adiabatic partition by one which is diathermic, or else by its complete removal.

Suppose on the contrary that S_C has decreased in a transition involving appropriate manipulation of partitions. Bearing in mind that all d-coordinates are freely variable, and that S_C remains constant when they are varied reversibly, we can arrange both S_A and S_B to have decreased. On the other hand there are certainly irreversible transitions such that S_A and S_B will both increase. In short, all states $\mathfrak{S}_C'$ [$\equiv (\mathfrak{S}_A', \mathfrak{S}_B')$] in a neighbourhood of any initial state $\mathfrak{S}_C$ are adiabatically accessible from $\mathfrak{S}_C$. This conclusion is, however, at variance with the existence of any (non-constant) entropy function of K_C at all. This conflict with the Second Law implies that our initial assumption was unjustified.

Note 24

Instead of concluding that such a process is impossible one can take the statement of its impossibility as an alternative formulation of the Second Law and hence proceed to define S and so on. This is, in fact, one of the traditional procedures based on the 'Principle of Kelvin'.

We must also bear in mind that the assumption that U has no upper bound is involved here, since without it we cannot conclude that S increases with U in isometric transitions. When the energy of a system does have an upper bound Kelvin's Principle may well be false. We shall return to this question at the end of the course. [See Problem **20.4**.]

Note 25

In chemistry a given constituent is denoted by some symbol, such as H_2O for water, H_2 and O_2 for gaseous (undissociated) hydrogen and oxygen respectively, and so on. The reaction in which hydrogen and oxygen combine to give water is then written $2H_2 + O_2 = 2H_2O$. Here we then more conveniently write C_1, C_2, C_3 in place of H_2, O_2 and H_2O and write the reaction equation in the form $-2C_1 - C_2 + 2C_3 = 0$. The equation of a general reaction is

$$\Sigma\, \nu_j\, C_j = 0,$$

where the *stoichiometric coefficients* ($= -2, -1, +2$ in our example) are taken as positive or negative according as the corresponding constituent appears or disappears in the reaction.

Note 26

We so avoid encumbering our equations needlessly. To accommodate more general situations we simply replace PV by $\Sigma\, P_k\, x_k$ in the definitions which follow, and likewise PdV by $\Sigma\, P_k\, dx_k$ and so on in subsequent equations. The only reservation is that we must ensure, by appropriate choice, that the x_k are extensive in the usual sense.

Note 27

In the physicist's notation a function such as the energy for instance is simply written as U, that is to say, without regard to the variables upon which U is taken to depend. If they are V, T one writes $U = U(V, T)$, if they are P, T one writes $U = U(P, T)$, and so on. This contains the seeds of ambiguity: and the mathematician would regard the indiscriminate use of U as a functional symbol as nonsense, writing, say, $U = f(V, T) = g(P, T)$ instead. Still, the physicist's notation has the great advantage of avoiding the endless proliferation of symbols associated with the same physical quantity. On the other hand, the formal relations of thermodynamics are then often disfigured by the attachment of subscripts to partial derivatives to indicate the variables which 'have to be kept constant'. One can avoid this in any given context by stating at the outset what independent variables are being used; and that is what we shall do here.

Note 28

We remind ourselves (recall also *Note* 19) that if $dL = Z_1dz_1 + Z_2dz_2 + \ldots + Z_ndz_n$ is to be a total differential of a function L, the *integrability conditions* $\partial Z_j/\partial z_k = \partial Z_k/\partial z_j$ $(j, k = 1, \ldots, n)$ must be satisfied.

Note 29

The conclusion that T cannot reverse its sign is not foolproof since it is contingent upon the assumption that any state of a given system can be reached *quasi-statically* from any other. Given some system for which this happens not to be the case the proof that $T''/T' > 0$ fails for all states $\mathfrak{S}''$ not quasi-statically accessible from a given state $\mathfrak{S}'$ and their absolute temperatures may well be opposite in sign to that of $\mathfrak{S}'$. We shall have occasion later on to consider this possibility in more detail.

We also see incidentally that only the ratios of absolute temperatures can be determined by experiment. To assign definite numerical values to the absolute temperatures of systems we must therefore first assign some specific, though freely disposable, value to the absolute temperature of some particular state of some particular system. The usual choice is the value 273.16 for the temperature of the triple point of water (see Lecture 10). This defines the *Kelvin scale* (°K).

Note 30

Any 'specific heat' C^* means the ratio dQ/dT, evaluated for an infinitesimal transition subject to $n - 1$ independent conditions which assure the uniqueness of this ratio. When these conditions are that all d-coordinates be kept constant we write C in place of C^*. In particular, when $n = 2$ and V is the d-coordinate, C is the specific heat at constant volume: $C = U_T$.

Note 31

The reason for the qualification 'formal' is this: to suppose that the integral corresponds to a realistic process entails the presupposition that the temperature $T = 0$ can actually be attained. Whether this is the case however remains to be seen.

Note 32

Only adiabatic cooling needs to be considered since otherwise we would have to have a system available whose temperature is already lower than that of the system of interest.

The occasional qualification that the reduction of the temperature to the absolute zero cannot be achieved 'in a finite number of operations' is to be rejected: first, because any asymptotic approach to $T = 0$ always leaves a non-zero temperature (and, as we have already discussed, who is to say that it is so 'small' as to be 'effectively zero'?), and in any event no-one can carry out an infinite number of operations.

Note 33

We now see the reason for contemplating only neighbouring pseudo-states. In any formal calculation designed to ascertain the details of a possible state of K (for example the internal state, given the external state) our rules compare values of the function S^* with each other. In general it might then well happen that a chosen set of values of the coordinates of K^* happens to be such that if K^* were actually in this state, removal of partitions etc. would lead to a state of K which is not the state of interest $\mathfrak{S}$. (As a matter of fact the state of K^* in question would then not be a pseudo-state corresponding to $\mathfrak{S}$ at all.) The restriction to infinitesimal variations to S^* prevents any such difficulty.

The situation is analogous to that of a billiard ball at rest at the bottom of a valley. In a small ('virtual') variation of its position its energy increases, but for large variations this need not be the case: there may be an adjacent, deeper valley. Likewise, if a thermodynamic system in equilibrium is subjected to a sufficiently large perturbation it may end up in a state different from the initial state.

Note 34

These masses are clearly independent of each other in the sense that the equilibrium conditions do not determine their relative values. Recall that the μ_j depend only upon the concentrations, so that the variation of the mass of any given phase with fixed concentrations cannot affect the equations $\mu_{\alpha j} = \lambda_j$.

Note 35

Many books on thermodynamics, some orientated towards physics, some towards chemistry, others again towards engineering are full of such applications. A selection from amongst those which I have found the most useful in the preparation of these lectures are the following:

Bazarov, I. P., (1966). *Thermodynamics.* Pergamon Press.

Buchdahl, H. A., (1966). *The Concepts of Classical Thermodynamics.* Cambridge University Press.

Callen, H. B., (1966). *Thermodynamics.* Wiley.

Falk, G. and Jung, H., (1959). *Axiomatik der Thermodynamik.* Article in Encyclopaedia of Physics, edited by Flügge, S., Vol. III/2, pp.119-175. Springer.

Huang, K., (1963). *Statistical Mechanics.* Wiley.

Kittel, C., (1958). *Elementary Statistical Physics.* Wiley.

Münster, A., (1959). *Prinzipien der statistischen Mechanik.* Article in Encyclopaedia of Physics, edited by S. Flügge, Vol. III/2, pp.176-412. Springer.

Pathria, R. K., (1972). *Statistical Mechanics.* Pergamon Press.
Schrödinger, E., (1948). *Statistical Thermodynamics.* Cambridge University Press.
Terletskii, Ya. P., (1971). *Statistical Physics.* North-Holland.
Tolman, R. C., (1950). *The Principles of Statistical Mechanics.* Oxford University Press.
Wilson, A. H., (1957). *Thermodynamics and Statistical Mechanics.* Cambridge University Press.

Note 36

By definition $\alpha = (\ln V)_T$, $\beta = (\ln P)_T$, $\kappa = -(\ln V)_P$, subscripts denoting partial derivatives, as usual. The equation of state $t = t(V, P)$ is supposed to have been solved alternatively for $V (= V(P, t))$ and $P (= P(V, t))$. Then the relation $V_P P_T = -V_T$ is merely a mathematical identity which is just the result in question written in terms of different symbols. (Which variables are kept constant in the various differentiations is obvious since the variables upon which the various functions are supposed to depend have been indicated explicitly.)

Note 37

By differentiation of G, $S = -G_T = \Sigma\, n_j s_j - R\, \Sigma\, n_j \ln c_j$. Suppose the gases to have been contained separately in individual compartments into which an adiabatic enclosure has been partitioned. Granted that P and T are uniform throughout this system, its entropy is $\Sigma\, n_j s_j$. Upon removing the partition the gases will diffuse into each other. Since overall Q and W vanish in this process, U, and therefore T, remain constant, as does P. It follows that the process of diffusion is accompanied by an increase of entropy $S = -R\, \Sigma\, n_j \ln c_j\, (> 0)$, explicitly confirming its irreversibility. (This result is usually established by a different argument, involving the reversible separation of the gases of a mixture, involving the use of semi-permeable membranes.)

Note 38

We have to be on our guard in supposing that an equation faithfully represents experimental results just because it seems to predict a few of them well enough. Thus consider the *critical point*, that is to say, that state $\mathfrak{S}_c$ for which $\partial P/\partial V$ and $\partial^2 P/\partial V^2$ both vanish. The van der Waals equation can accommodate this situation, and it predicts that the ratio PV/RT should have the value $\frac{3}{8}$ at $\mathfrak{S}_c$. This compares not unfavourably with the values of the order of 0.29 found for common gases. Now consider instead the virial expansion, arbitrarily truncated after the third term. This time PV/RT turns out to have the value $\frac{1}{3}$ at $\mathfrak{S}_c$ which is better than that provided by van der Waals' equation. Yet the truncation of the series is hardly justified in the proximity of $\mathfrak{S}_c$, where the behaviour of the gas is surely very far from ideal. The fact is that the value of the ratio PV/RT at $\mathfrak{S}_c$ predicted by the equation of state seems to be remarkably insensitive to its precise form, and the extent to which it agrees with experiment provides no effective criterion for deciding whether a given equation of state is likely to be satisfactory or not.

Note 39

The fact that C_i does not remain even approximately an ideal gas at low temperatures is irrelevant here, granted that the integral makes allowance for heats of evaporation, of melting, and the like.

Note 40

Of course, in other branches of physics one also often compares a given state of affairs with some fictitious condition, namely, whenever laws are formulated as variational principles (Fermat's Principle, Hamilton's Principle, and the like). The question just raised is equally relevant to these, but it is mostly ignored. In fact, nature is not as simple as such variational principles might suggest, and they always occur in theories which are approximations to more complete, and conceptually distinct, theories. Non-relativistic quantum mechanics, for instance, replaces classical (Newtonian) mechanics and its formulation based on the idea of 'sums over histories' serves as a picturesque reminder of the way in which a dynamical system explores those motions which are forbidden by Hamilton's Principle.

Note 41

If the conditions envisaged here are not fulfilled we get into inevitable difficulties in that we shall be confronted with problems which, with the information actually available, are indeterminate. As a typical example, take K to be once again our sample system and suppose it to be adiabatically isolated from $K^\dagger$. The latter is here taken as a gas literally surrounding K and itself enclosed in a rigid adiabatic envelope. Let the piston, which is part of the enclosure of K, be initially at rest, being held in position by the equal and opposite forces exerted on it by K and $K^\dagger$. Upon displacing the piston and then releasing it, it will oscillate to and fro, the oscillations gradually dying out on account of frictional forces within both K and $K^\dagger$. However, since we know nothing about the nature of these frictional forces the final position of the piston is unpredictable. When the piston is diathermic the situation is quite different, for we then have two additional pieces of information: both initially and finally the temperature of K must be equal to the temperature of $K^\dagger$, and the problem is no longer indeterminate.

Note 42

K_1, K_2 and K have representative ensembles $\mathscr{K}_1$, $\mathscr{K}_2$ and $\mathscr{K}$ respectively. In the present context we conveniently take $\mathscr{K}_1$ and $\mathscr{K}_2$ to be the ensembles consisting of all those members of $\mathscr{K}$ which remain when $\mathscr{K}_2$ and $\mathscr{K}_1$, respectively, are ignored throughout.

Note 43

Differentiating the given relation with respect to σ_1 and σ_2 respectively, it follows immediately that $p_1'/p_1 = p_2'/p_2$, where primes denote differentiation with respect to the argument. Since σ_1 and σ_2 are independent variables each ratio must have the same constant value, $-1/k$, say. Thus $p_j'/p_j = -1/k$ ($j = 1, 2$) whence $p_j = \kappa_j \exp(-\sigma_j/k)$, where κ_j is a constant of integration. Since we must have $p_j > 0$ and $p_j' < 0$, k must be positive.

Note 44

Of course it might happen that there are no quadratic terms. Cubic terms alone could not be admitted and at least the quartic terms would have to be included to ensure convergence of the integral $\int \exp(-\sigma/k)da_1 da_2 \ldots$. However, the present simple-minded phenomenological theory is in any case inadequate in such situations. This is true for instance at a critical point, where the dominant terms would indeed be quartic, but terms of a different kind altogether have been disregarded.

Note 45

To evaluate J we first orthonormalize the quadratic form $\Sigma\, a_{ij}\, a_i\, a_j$, that is, we make a linear transformation $a_i = \Sigma\, L_{ij}\, b_j$—or, in matrix notation, $a = Lb$—such that $\tilde{a}aa = \tilde{b}b$, where a tilde denotes the transpose. Therefore we must have $\tilde{L}aL = 1$, and on taking the determinant of both sides $(det\ L)^2\ (det\ a) = 1$, i.e. $det\ L = j^{-\frac{1}{2}}$. The transformation changes J into $(det\ L) \int \exp(-\Sigma\, b_i^2)db_1db_2 \ldots = j^{-\frac{1}{2}} \{\int \exp(-b^2)db\}^r$. The only significant contribution to J comes from a small neighbourhood about $\sigma = 0$ and the integration may safely be extended from $-\infty$ to ∞. The required result $J = (\pi^r/j)^{\frac{1}{2}}$ thus follows at once.

Note 46

This is not merely to say that every theory has to be validated by experiment. The point which must not be overlooked here is that the microscopic equations of motion are taken to be valid. Even then, however, one is always confronted with difficulties concerning the equivalence of temporal and ensemble averages, and also with difficulties surrounding the contrast between macroscopic irreversibility and microscopic reversibility.

Note 47

This is all part and parcel of the elementary theory of analytical mechanics. The q_k are 'generalized coordinates' the dimensions of which are unspecified, which explains the quotation marks around 'velocities' and 'accelerations'. Note, however, that the dimensions of p_kq_k, i.e. ml^2t^{-1}, are independent of those of q_k. If all forces are derivable from a potential, L is simply the kinetic energy minus the potential energy of K as a whole. If magnetic forces are present one has to add a term bilinear in the velocities and the components of the magnetic vector potential. Anyone quite unfamiliar with the subject may wish to consult a book such as that by H. Goldstein, (1953), *Classical Mechanics*. Addison-Wesley.

Note 48

Suppose one goes over from one set of variables to another, p^*, q^* say, such that the transformed equations of motion are again Hamiltonian, i.e. $\dot{q}_k^* = \partial H^*/\partial p_k^*$, $\dot{p}_k^* = -\partial H^*/\partial q_k^*$. This change may be made, for instance, by choosing initially a different set of coordinates, and then finding the p_k^* and H^* from L^* by the usual prescription; or else such a transformation might be achieved in a more general way. Then the Jacobian of any transformation of this kind is unity. This entails that a multiple integral of the kind $\int f(p, q)dp_1dq_1dp_2 \ldots dq_r$ transforms simply into $\int f(p, q)dp_1^*dq_1^* \ldots dq_r^*$. In other words, in this context all sets of variables p, q in terms of which the equations of motion take the Hamiltonian form can be formally treated as if they were Cartesian coordinates.

Note 49

To establish this result it is helpful to think of the phase of any given member of $\mathscr{K}$ to be pictured by a point in a $2r$-dimensional representative space Γ, called the *phase space* of K. The phase point moves about in Γ as K pursues its history, so that $\mathscr{K}$ as a whole is represented by a moving cloud of phase points. Since the number ν of such phase points is very large ($\nu \rightarrow \infty$), the cloud can be approximated by a continuous 'fluid' of density

$\rho = v\varphi$. Moreover, since no phase point can disappear from Γ and its motion is continuous, $-\partial\rho/\partial t$ must, as usual, be equal to the divergence of the 'current' whose components are $\rho\dot{p}_1, \rho\dot{q}_1, \ldots, \rho\dot{q}_r$. Bearing in mind that the coordinates of Γ can be treated as Cartesian [see *Note* 48] this divergence is $\Sigma[\partial(\rho\dot{p}_k)/\partial p_k + \partial(\rho\dot{q}_k)/\partial q_k]$. Because of the equations of motion $\partial\dot{q}_k/\partial q_k + \partial\dot{p}_k/\partial p_k = 0$ and the required conclusion follows at once.

Note 50

This result is evidently crucial in the context of the so-called 'postulate of equal *a priori* probabilities' which forms part of the foundations of the subject when they are treated in a more axiomatic spirit. This postulate assigns equal *a priori* probabilities for different regions of phase space of equal extent. (In effect we used this postulate when we defined the quasi-uniform ensemble.)

Note 51

We might wonder about the presence of a derivative on the left. Now, the possible appearance of a constant of integration is equivalent to the formal possibility of replacing w by $w^* +$ const. φ^{-1}. However, the existence of $\overline{w}$ precludes the existence of $\overline{w^*}$, so that the constant must be zero.

Note 52

The inconsistency of these results with the earlier equation $(\varphi w)' = H/T - \Lambda$ is only apparent. On the left we now have $-k[1 + \ln(\varphi/a)] = w - k$ whilst on the right we have w. However, in the steps following this equation we effectively make use only of the variation of $H/T - \Lambda$, and $\delta k \equiv 0$, of course.

Note 53

It suffices to consider an ideal gas, which has $F = -kT(\ln a + N\ln V + N\tau(T))$. Extensivity requires that $F = Nf(V/N, T)$, so that we must have $N^{-1}\ln a = -N\ln N + N\ln\gamma$, where γ is some constant.

Strictly speaking extensivity obtains only in the *thermodynamic limit*, in which N and V tend to infinity together in such a way that V/N tends to a finite limit v. Hence N must here be very large. It is important to realize that the appeal to extensivity for the purpose of determining a, and so of the dependence of S upon N, is essentially a *convention* in the context of the canonical ensemble. Thermodynamically only the entropy differences between states which are reversibly accessible from each other are defined, so that, N being fixed, nothing can be said about the dependence of S on N in the first place.

Note 54

This (non-relativistic) relation between $\mathbf{u}$ and the canonical momentum $\mathbf{p}$ is not valid when the particle carries an electric charge e and is acted upon by an electromagnetic field, for then $\mathbf{u} = \mathbf{p}/m - (e/mc)\mathbf{A}$, where $\mathbf{A}$ is the vector potential. However, it is still true that $d\mathbf{u}$ and $d\mathbf{p}$ differ only by a constant factor.

Note 55

We have generally $Z\overline{N^s} = \Sigma\, a_N\, N^s\, \lambda^N\, Z_N$, $Z = \Sigma\, a_N\, \lambda^N\, Z_N$. Therefore, by inspection, $\overline{N} = kT\, \partial Z/\partial\mu$, $\overline{N^2} = (kT)^2\, \partial^2 Z/\partial\mu^2$, and then $(\delta N)^2 = (kT)^2\, \partial^2 \ln Z/\partial\mu^2 = kT\, \partial\overline{N}/\partial\mu$.

The Gibbs-Duhem identity, now written in terms of $\overline{N}$, rather than n, is

$$\overline{N}\, d\mu = -S\, dT + V\, dP,$$

so that, bearing in mind that P and μ are functions of $V/\overline{N}$ and T, $V\partial P/\partial V = \overline{N}\partial\mu/\partial V = -\overline{N}^2\, V^{-1}\, \partial\mu/\partial\overline{N}$. It follows that

$$(\delta N/\overline{N})^2 = -kT\, V^{-2}\, \partial V/\partial P = kT\, \kappa/V.$$

For an ideal gas $\kappa = 1/P$ and so $\delta N/\overline{N} = \overline{N}^{-\frac{1}{2}}$.

Note 56

Pathological behaviour is characterized by, or 'caused by', discontinuities of the thermodynamic potentials. For example, formally $\delta N \to \infty$ when $\kappa \to \infty$, i.e. when $\partial^2 G/\partial P^2 \to \infty$. (This happens for instance as one approaches the triple point of a van der Waals gas.) Clearly arguments concerning the equivalence of the description of systems afforded by generically different ensembles do not apply in pathological situations.

Note 57

If we have a substance which is not ideal we can first consider it under conditions under which it is effectively ideal (V and T sufficiently large). a_N^* however does not depend on V and T, so that its value must still be that which we are about to derive, whatever the state of the substance may be.

Note 58

In fact we are simultaneously confronted with two kinds of ensembles, or better, an ensemble of ensembles. The reason for this is that in the quantum mechanical description of a simple system K which is in some (normalized) state $|\rangle$ a single measurement of an observable $\hat{A}$ will not in general yield a value which can be predicted with certainty. On the contrary, the outcome of the experiment must be one of the eigenvalues A', A'', ... of $\hat{A}$, a particular value, A' say, being obtained with the probability $|\langle A'|\rangle|^2$, where $|A'\rangle$ is the eigenstate of $\hat{A}$ belonging to A'. Only if $|\rangle$ happens to be $|A'\rangle$ will the value A' result with certainty. In other cases 'the value of $\hat{A}$' is understood to be the mean of the values obtained in simultaneous measurements made on a large number of copies of K, all prepared in the same way to within the limitations imposed by the principles of quantum mechanics. Evidently this *expectation value* $\langle A\rangle$ of $\hat{A}$ ($\equiv \langle|\hat{A}|\rangle$) is itself an ensemble mean. The 'thermodynamic value' of $\hat{A}$ is then $\overline{\langle A\rangle}$.

Note 59

Let a randomly selected member K_a of $\mathscr{K}$ be in a state $|n\rangle$ which is one of a discrete orthonormal set. The expectation value of some observable $\hat{A}$ is then $\langle n|\hat{A}|n\rangle$. Let P_n be the probability that K_a is in fact in the state $|n\rangle$. Then the 'thermodynamic value' A of $\hat{A}$ is $\Sigma\, P_n\, \langle n|\hat{A}|n\rangle$ (recall the preceding *Note*). If $|v\rangle$ is an arbitrarily chosen representation we therefore have $A = \Sigma\Sigma\, P_n\, \langle n|\hat{A}|v\rangle\langle v|n\rangle$, or $A = \sum_v \langle v|\rho\hat{A}|v\rangle$, where $\rho = \sum_n |n\rangle\, P_n\, \langle n|$.

ρ is the *statistical operator*. (Note that the Schrödinger representative of ρ is $\rho_{pq} = \sum_n \langle E_p | n \rangle P_n \langle n | E_q \rangle = \overline{\psi_q^* \psi_p}$.)By inspection $A = tr(\rho \hat{A})$, this trace being independent of the representation. In particular $tr\, \rho = 1$, which is the analogue of the condition $\int \varphi \, d\Gamma = 1$.

If the states $|n\rangle$ are eigenstates of H (i.e. $|n\rangle \equiv |E_n\rangle$) then

$$A = \Sigma \langle E_n | \rho \hat{A} | E_n \rangle = \Sigma A_n \langle E_n | \rho | E_n \rangle = \Sigma A_n \varphi_n,$$

say, and if ρ is merely a function of H, $\varphi_n = \rho(E_n)$. Incidentally, one has here also an analogue of the postulate of equal *a priori* probabilities (see *Note* 50), but we need not go into this.

Note 60

The equation is $\nabla^2 \psi + 8\pi^2 m h^{-2} E \psi = 0$, and ψ must vanish when $x = 0$ and when $x = l (= V^{\frac{1}{3}})$, and likewise when y and when z take one or other of these values. Therefore $\psi = \text{const.} \times \sin(\pi a x/l) \sin(\pi b y/l) \sin(\pi c z/l)$ is an admissible solution provided a, b, c are positive integers. The negative integers would represent the same *states* and must therefore not be counted. By substitution in the equation, $8mh^{-2} E = (a^2 + b^2 + c^2)/l^2$. (Here we have of course only considered the non-relativistic case.)

Note 61

The particles cannot be physically distinguished from each other and their formal interchange must leave the state, or more precisely $|\psi|^2$, unaffected. Thus, under the mutual interchange of the ith and jth particles ψ must go into $\lambda_{ij}\psi$, where λ_{ij} is a phase factor. Interchanging them again, we recover the same wave function, so that λ_{ij}^2 must be unity. It is not difficult to show that $\lambda_{ij} = \lambda_{kl}$ $(i \neq j, k \neq l)$ by permuting two of a set of three particles at a time. It follows that for a given system either $\lambda_{ij} = 1$ for all i and j $(i \neq j)$ or else $\lambda_{ij} = -1$.

Note 62

To show that $\delta E_j = \langle \delta H \rangle_j$, we consider the effect of a variation of the d-coordinates on the eigenvalue equation $(H - E_j)|j\rangle = 0$. Thus $[(\delta H) - \delta E_j]|j\rangle + (H - E_j)(\delta |j\rangle) = 0$. Forming the product with $\langle j |$ throughout, $\langle j | (\delta H) - \delta E_j | j \rangle = 0$ since $\langle j | (H - E_j) = 0$, so that we already have the required result.

Note 63

We compute the Helmholtz function for a non-canonical distribution $\varphi_j^* = \varphi_j e^{\eta_j}$, where the η_j are arbitrarily chosen numbers. Thus $F^* = \Sigma[E_j + kT \ln (\varphi_j^*/\gamma)]\varphi_j^*$, which becomes $\triangle F = kT \Sigma\, \eta_j e^{\eta_j} \varphi_j$. Since $\Sigma \varphi_j e^{\eta_j} = \Sigma \varphi_j = 1$ we can write

$$\triangle F = kT \Sigma[(\eta_j - 1)e^{\eta_j} + 1]\varphi_j$$

and $\triangle F > 0$ unless $\eta_j = 0$ (all j).

Note 64

If $\psi_i(q_k)$ is a one-particle wave function, q_k being nominally the coordinates of the kth particle, the wave function of K is

$$\psi(q_1, q_2, \ldots, q_N) = \Sigma\, \eta_P\, P(\psi_{i_1}(q_1)\, \psi_{i_2}(q_2) \ldots \psi_{i_N}(q_N)),$$

where the sum extends over all the permutations of the i_r, whilst $\eta_P = 1$ in the symmetric case and equal to the sign of the permutation in the antisymmetric case.

Note 65

I is the moment of inertia about a line through the centre of mass and perpendicular to the line joining the atoms. If their masses are m, m', $I = mm'r_0^2/(m + m')$, where r_0 is to be taken as the equilibrium distance between the atoms.

Note 66

If the spin wave functions of the first nucleus are $\psi_\sigma(1)$ and those of the second are $\psi_\tau(2)$ ($\sigma, \tau = 1, \ldots, \rho = 2s + 1$), there are $\frac{1}{2}\rho\,(\rho + 1)$ composite symmetric wave functions, namely $\frac{1}{2}\rho\,(\rho - 1)$ of the form $\psi_\sigma(1)\,\psi_\tau(2) + \psi_\sigma(2)\,\psi_\tau(1)$, $\sigma \neq \tau$, and ρ of the form $\psi_\sigma(1)\,\psi_\sigma(2)$; and there are $\frac{1}{2}\rho\,(\rho - 1)$ antisymmetric wave functions of the form $\psi_\sigma(1)\,\psi_\tau(2) - \psi_\sigma(2)\,\psi_\tau(1)$.

Note 67

It will suffice to consider H_2. Then $z = \zeta_+ + 3\zeta_-$, so that the ratio of the number of molecules with odd j to that with even j is $\rho_0 = 3\zeta_-/\zeta_+$. When $T \gg T_r$, $\rho_0 = 3$, whereas ρ_0 goes to zero as $T \to 0$. Now in any measurement of the specific heat of H_2 one will normally start with a sample of the gas at room temperature and then cool it down to T_r and below. The establishment of equilibrium would require ρ_0 to attain the value appropriate to the temperatures at which measurements are made. This adjustment involves transitions between different nuclear spin states, a process whose time scale (in the absence of a catalyst) is of the order of months rather than minutes. In the course of an experiment ρ_0 will therefore not have its equilibrium value but the value corresponding to room temperature; in effect $\rho_0 = 3$. To all intents and purposes the gas therefore behaves as a mixture of two distinct constituents, the first with j even *(para-hydrogen)* and the second, three times as abundant, with j odd *(ortho-hydrogen)*. The final equation for C_r (that is, C_{r-} with $s = \frac{1}{2}$ here) clearly corresponds to this situation.

Note 68

One can also contemplate 'ν-parafermions', that is, particles such that any given level can be occupied at most ν times. (Whether they exist in nature may be left aside.) Then $Z = \prod_i [(1 - \eta_i^{\,\nu+1})/(1 - \eta_i)]$ and this includes the fermion case ($\nu = 1$) and the boson case ($\nu = \infty$).

Note 69

Whilst investigating the classical limit we found that $Z_N = (\omega V l^{-3})^N/N!$, so that $\ln Z = \lambda\omega V/l^3 = PV/kT = \overline{N}$. Therefore $\lambda = \overline{N}l^3/\omega V$.

Note 70

By inspection $\eta_k\,\partial Z/\partial\eta_k = \Sigma\,\Sigma\Pi\, n_{jk}\,\eta_i^{\,n_{ji}} = Z\,\overline{n}_k$, and a second differentiation gives $\eta_k\partial(\eta_k\,\partial Z/\partial\,\eta_k) = Z\,\overline{n_k^2}$. Thus $\overline{n}_k = \partial \ln Z/\partial\eta_k$ and

$$(\delta n_k)^2 = \overline{n_k^2} - \overline{n}_k^2 = \eta_k\,\partial(\eta_k\,\partial \ln Z/\partial\eta_k)/\partial\eta_k.$$

Inserting $\ln Z = \pm\,\Sigma \ln(1 \pm \eta_i)$, the stated results follow immediately.

Note 71

In the integral $J = \int_0^\infty (e^{t-\tau} + 1)^{-1} \varphi(t)dt$, $(\tau \gg 1)$, split the range of integration at $t = \tau$. Then $J = \int_0^\tau \varphi(t)dt - \int_0^\tau (1 + e^{\tau - t})^{-1} \varphi(t)dt + \int_\tau^\infty (1 + e^{t-\tau})^{-1} \varphi(t)dt$. Put $\tau - t = s$ in the second integral and $t - \tau = s$ in the third: $J = \int_0^\tau \varphi(t)dt - \int_0^\tau (1 + e^s)^{-1} \varphi(\tau - s)ds + \int_0^\infty (1 + e^s)^{-1} \varphi(\tau + s)ds$. We can extend the second integral to ∞ with insignificant error, granted that φ is a slowly varying function of t (e.g. a polynomial). By the same token

$$
\begin{aligned}
J &= \int_0^\tau \varphi(t)dt + \int_0^\infty \{2\varphi'(\tau)s + \varphi'''(\tau)s^3 + \ldots\}(1 + e^s)^{-1} ds \\
&= \int_0^\tau \varphi(t)dt + 2\varphi'(\tau)\int_0^\infty s(1 + e^s)^{-1} ds + \ldots \\
&= \int_0^\tau \varphi(t)dt + \frac{\pi^2}{6}\varphi'(\tau) + \ldots .
\end{aligned}
$$

Note 72

For disordered, i.e. isotropic, radiation of any kind $U \doteq 3PV$, and thermodynamic arguments shows that the energy density $u = U/V$ of black body radiation depends on T only. Now $TdS = dU + (U/3V)dV = \frac{4}{3}udV + Vdu = V^{-\frac{1}{3}}d(uV^{\frac{4}{3}})$, by inspection. Hence $TV^{\frac{1}{3}}$ must be a function of $uV^{\frac{4}{3}}$ alone, and since u does not depend on V, it follows that $u = \sigma T^4$, where σ is a constant. Hence $S = \frac{4}{3}\sigma T^3 V$ and so $G = U + PV - TS = 0$.

Note 73

In typical experiments the kind of times with which one is confronted are of the order of microseconds for θ_m and minutes for θ_l. One is then fairly safe in experiments carried out in times θ of the order of a few seconds. To account for what is actually observed requires a considerable theoretical apparatus and a large number of complicating factors have to be taken into account; see, for example, the paper by A. Abragam and W. G. Proctor, *Phys. Rev.*, **109** (1958), 1441-1458.

Note 74

Of course that cannot be done. What is meant is that the reversal should take place in a time much less than θ_m. This time being in fact finite, the spin system will not be in equilibrium immediately after reversal but only a time of the order of θ_m later. This brings with it that results such as $S' = S''$ which we are about to encounter are only approximately valid. We can safely disregard this complication.

Note 75

This is subject to the qualification of the preceding *Note*. The reason for raising this point again is that the case $j = \frac{1}{2}$ is exceptional. That is why we earlier calculated S and U for general values of j. When $j = \frac{1}{2}$ the population numbers $N_{+\frac{1}{2}}$ and $N_{-\frac{1}{2}}$ can never change as a consequence of spin-spin interactions alone since the spin flip $m = -\frac{1}{2} \rightarrow \frac{1}{2}$ of one nucleus must be accompanied by the spin flip $m = \frac{1}{2} \rightarrow -\frac{1}{2}$ of another, if energy is to be conserved. Whatever $N_{+\frac{1}{2}}$ and $N_{-\frac{1}{2}}$ may happen to be, the temperature is in effect defined by the relation $N_{+\frac{1}{2}}/N_{-\frac{1}{2}} = \exp(\gamma B/kT)$, so that the situation is rather trivial. When $j \neq \frac{1}{2}$ it is otherwise, since then the simultaneous spin transition $m = m' \rightarrow m' + 1$ and $m = m' \rightarrow m' - 1$ for example $(m' \neq j)$ conserve energy, granted that the various levels are (as here) equidistant.

SOLUTIONS OF PROBLEMS

1.1 A physical quantity is defined by a well-defined sequence of operations and calculations whose outcome is a number which is the value of the quantity under the given circumstances. In other words, the phrase 'such and-such a physical quantity' is a kind of shorthand for a properly defined experimental and computational routine. In ascribing a 'distance' to an object, a measurement of this quantity might involve either measuring rods or else the use of radar, to take just two possibilities. Clearly one really has then *two* physical quantities, i.e. 'rod-distance' and 'radar-distance', not just 'distance'. This is as should be, for there is no *a priori* reason why rod-distance and radar-distance should be equal. We conclude that in the case of entropy we certainly have two distinct physical quantities, i.e. phenomenological entropy and statistical entropy (each separately defined by its own routine for obtaining its value).
1.2 No. It is equally impossible to test by direct experiment whether a given system in fact satisfies the condition in question.

2.1 Any monotonic function of P, e.g. $1/P$, will do. More generally, any function of P and V, monotonic in both (PV for example) may be adopted.
2.2 No. Iron exhibits (magnetic) hysteresis effects, i.e. its properties at any given time depend on its past history.
2.3 (ii) and under special circumstances (iii). As regards (i), leaving aside that it is not in equilibrium (the moon will 'cool down' in the course of time) self-gravitation cannot be disregarded. As for (iii) the question is (deliberately) somewhat ambiguous. The gas might be water vapour in which case it is a standard system. On the other hand, if we start with a mixture of oxygen and hydrogen under normal conditions, then adiabatically changing the volume and restoring it again to its initial value, we may end up with the same mixture or we may not. For instance if the mixture was sufficiently compressed at some stage water vapour can have formed by chemical reaction. Thus the condition of independence of previous history is violated. (As a matter of fact this conclusion merely reflects that *on a sufficiently long time scale* an oxygen-hydrogen mixture is not in equilibrium so that, strictly speaking, the quantitative description of this system is beyond the competence of equilibrium thermodynamics.) Finally, (iv) involves non-negligible surface tension.

3.1 In case (i) there is a well-defined force which the gas exerts on the piston, and this force may here be taken for granted to depend upon the volume alone. Hence the work done by the gas during the increase in volume is exactly recovered by the surroundings during the second stage of the process. In case (ii), on the other hand, the gas does no work at all in the first stage. [Somewhat naïvely expressed, because of its inertia it cannot follow the motion of the piston and so exerts no pressure on it.] In the second stage, the surroundings must again do work on the gas. In short, in case (ii), but not in case (i), the final condition of the external device is necessarily different from its initial condition, the change reflecting the net amount of work done on the gas. (At the same time we find that the final pressure of the gas in case (ii) is greater than the initial pressure.)

3.2 Mechanical. For simplicity we here take the field to be due to permanent magnets and to change the field they have to be moved about.

3.3 Unless the stirrer turns at an infinitesimal rate the question is nonsense since the system is not in a state. (In the pseudo-static case $n = 2$.)

3.4 Yes, but only in a very artificial sense. An apparatus incorporating a ratchet mechanism is an example. Of course, we disregard such 'trivial' sources of irreversibility throughout.

3.5 Because it has two h-coordinates.

4.1 All states $\mathfrak{S}_2\,[P_1, V_2]$ such that $P_2V_2^{\gamma} < P_1V_1^{\gamma}$. We can always first adjust the volume reversibly to its desired value V_2 and the pressure is then $P_1\,(V_1/V_2)^{\gamma}$. This can only be changed to P_2 (say by stirring) if it is less than P_2.

4.2 No. In fact, given some s subject to the original inequality, one need only reverse its sign. Of course, subsequent conclusions would have to take the reversal of the inequality into account. For example, the Entropy Principle would become a 'Principle of Decrease of Entropy'.

4.3 The formulation does not rigorously imply that the function s is not a constant (for all values of the coordinates). If s were a constant the ordering would be trivial and all states of K_0 would be mutually accessible (cf. *Note* 9). That s is not in fact a constant function should therefore really be stated explicitly.

5.1 No. No net work is being done by the earth in the course of any one revolution. If one tried continually to extract work the mean distance from the sun would continue to decrease as revolution follows upon revolution.

5.2 $\triangle U = 0$ since $Q = 0$ and $W = 0$.

5.3 $s = U$, or more generally $s = f(U)$, where f is any good monotonically increasing function. In turning the stirrer work has to be done on the system and so one can only have transitions to states of greater energy. Any such transition is at the same time irreversible.

6.1 A system K^* which is standard except for the presence of an internal adiabatic partition. A thermometer brought into contact with one part of K^* may be in equilibrium with it, but may not be so when brought into contact with another part of K^*.

6.2 The sun-thermometer system is not in equilibrium—not even approximately. Characteristically two ordinary mercury-in-glass thermometers, one with the bulb painted white and the other black, but both otherwise identically constructed, will give quite different readings when placed side by side in the sun, even more so if the incident light is focused upon one of the bulbs by means of a magnifying glass.

6.3 Granted that the water is, at least approximately, at atmospheric pressure throughout, it attains its greatest density at about 4 °C. It follows that t is not a single-valued function of V in a neighbourhood of $t = 4$ °C, i.e. it is not a good function. In fact, in the range $0 < t < 8$ (°C) one has, very roughly, $V = 1 + 8 \times 10^{-6}\,(t - 4)^2$, there being just one gram of water; and here the double-valuedness of t is manifest.

6.4 In equilibrium the temperature has the same value everywhere. In the absence of equilibrium the temperature is not defined, so that there is no sense in an equation purporting to describe the 'distribution of temperature'. (Of course, this must not be taken too seriously. Pragmatically one can introduce the notion of 'local equilibrium' which remains meaningful if the value of the temperature, now defined operationally in terms of specific thermometers, turn out to be independent of the nature of these to within experimental error; cf. *Note* 4.)

7.1 $dS = (dU + PdV)/T$

$$= [(\beta + 1)t/TV]dV + [(a + \beta \ln V)/T]dt = AdV + Bdt,$$

say. Since dS is a total differential, the t-derivative of A must be equal to the V-derivative of B, i.e. $(\beta + 1)tdT/dt = T$, whence the required result follows at once.

7.2 Yes, granted that we can take K and $\bar{K}$ jointly to constitute a composite adiabatically isolated system K^*. In the adiabatic transition S^* increases because S does so, and in the subsequent transition S^* cannot decrease. Hence S^* has increased overall, and K being again in its initial state the same cannot be true of $\bar{K}$. (This argument presupposes that the surroundings constitute a thermodynamic system for which an entropy function is defined. This may seem artificial at first sight, but it corresponds to practical realities. How else, in general, is one to say whether 'changes' in $\bar{K}$ have occurred, or, if they visibly have, whether they are relevant?

7.3 Entropy. (It is undesirable, however, to rephrase the conclusion that entropy can only be created by saying that 'the entropy of the universe is increasing', since few, if any, of our basic concepts are defined, or even definable, for systems of cosmical extent.)

7.4 Since $dS = (dU + PdV)/T = [(P + \partial U/\partial V)dV + (\partial U/\partial T)dT]/T$ is a total differential, we must have (see Problem **7.1**) $\partial U/\partial V = T(\partial P/\partial T) - P$. With $P = RT/V$, this shows that U is a function of T alone. Therefore $dS = T^{-1}(\partial U/\partial T)dT + RdV/V$, whence

$$S = \int_{T_1}^{T} T^{-1}(\partial U/\partial T)dT + R \ln V + S_1,$$

where T_1 is an arbitrarily chosen temperature and S_1 a constant of integration.

8.1 Yes. Write $dU = TdS - PdV + \Sigma \mu_j dn_j$, and we see at once that the appropriate variables are S, V and the n_j.

8.2 No. Amongst the functions to be determined is $V(P, T)$ in particular. However, now $dF = -PdV - SdT = -[P(\partial V/\partial T) + S]dT - P(\partial V/\partial P)dP$, and so V has to be obtained from the differential equation $P\partial V/\partial P = -\partial F/\partial P$ and this determines it only to within an arbitrary additive function of T. An analogous result holds for S. In short, $F(T, P)$ does not uniquely specify the various basic thermodynamic functions. The crucial point is that almost any quantity such as P or V or S becomes a characteristic function in appropriate circumstances once the variables on which they are taken to depend are properly chosen.

8.3 To say that P is independent of V means that in the context of the particular coordinates which occur here P does not depend *explicitly* upon V. Of course it depends on μ, so that there is no contradiction. Consider the specific example $X = a\, T^{\beta} e^{\gamma\mu/T} V$, where a, β and γ are constants. Then $\partial X/\partial V = P$ does not involve V. On the other hand, $n = \partial X/\partial \mu = \gamma X/T$ here, i.e. $X = PV = n\gamma T$, which is just the equation of state of a classical ideal gas.

8.4 One doesn't need to obtain it: $P + \partial F/\partial V = 0$ is already the equation of state.
8.5 $G = U - TS + PV$ and, if u is the energy per mole, we have from Problem **7.4** $G = nu(T) - nT\{\int_{T_1}^{T} T^{-1}(\partial u/\partial T)dT + R\ln(V/n) + S_1\} + nRT$, where S_1 is now reckoned per mole. Here V must still be eliminated in favour of P ($= nRT/V$). Then, by inspection

$$\chi(T) = \int_{T_1}^{T} c\, dT - T\int_{T_1}^{T}(c/T)dT - RT\ln T + a_1 T + a_2,$$

where $c = \partial u/\partial T$ and a_1, a_2 are constants. If we write $c = \hat{c} - R$,

$$\chi(T) = \int_{T_1}^{T} \hat{c}\, dT - T\int_{T_1}^{T}(\hat{c}/T)dT - \beta T + \gamma,$$

where β and γ are constants.

9.1 At once $S_V = P_T = R/V$ and so S certainly cannot become independent of V as $T \to 0$. The existence of such a substance would therefore be in conflict with the Third Law.
9.2 No. From the identity $U_V = TP_T - P$ already encountered we infer that $TP_T = 4P$, i.e. $P = bT^4$, where b is a constant. Thus $U = 3bVT^4$, whence $S = 4bT^3V$, an irrelevant additive constant having been omitted. Thus S_V vanishes with T for all values of V.
9.3 Proceeding as in the preceding problem we at once get the differential equation $3VP_V - 2TP_T + 5P = 0$. Introduce the function $Y = PV/T$ in place of P, and then $3VY_V = 2TY_T$, i.e. Y is some function of $TV^{\frac{2}{3}}$. Next, using the equations $PV = \frac{2}{3}U = Tg(z)$ in the usual relation $TdS = dU + PdV$, we find that $S = \frac{3}{2}\int(g' + z^{-1}g)dz$, with $g' = dg/dz$. From this $C = TS_T = zS_z = \frac{3}{2}(zg' + g)$, so that $S = S_1 + \int_{z_1}^{z} z^{-1}\, Cdz$, where z_1, S_1 are constants. That C goes to zero with T as T^n means that $C \sim az^n$ as $z \to 0$, where a is a constant. Therefore the integral for $S - S_1$ converges as $z \to 0$, i.e. S is a constant independent of V. On the other hand, if $C \sim a/\ln z$ for instance, the integral does not converge since $\int(z\ln z)^{-1}\, dz = \ln\ln z$, even though $C \to 0$ with T.

10.1 Yes. Our previous conclusion that $f = 3 - \varphi$ followed under the assumption that $n = 2$. However, if, for example, the constituent is magnetic, its magnetic moment in the direction of an external field is a second d-coordinate and so $n = 3$ and therefore $f = 4 - \varphi$.
10.2 We have $S^* = S_A + S_B$. Hence, conveniently taking U_A, U_B as variables in place of T_A, T_B (V_A and V_B being fixed)

$$S^* - S = (\dot{S}_A\, dU_A + \dot{S}_B\, dU_B) + \tfrac{1}{2}[\ddot{S}_A\,(dU_A)^2 + \ddot{S}_B\,(dU_B)^2] + \ldots,$$

dots denoting derivatives with respect to energy (at constant volume). Now $\dot{S} = T^{-1}$ and $\ddot{S} = \partial(T^{-1})/\partial U = -C^{-1}T^{-2}$, so that, since $dU_B = -dU_A$,

$$S^* - S = (T_A^{-1} - T_B^{-1})dU_A - \tfrac{1}{2}(C_A^{-1}T_A^{-2} + C_B^{-1}T_B^{-2})(dU_A)^2 + \ldots.$$

The condition $dS^* = 0$ of course merely gives $T_A = T_B$ ($= T$, say), and then

$$d^2S^* = -\tfrac{1}{2}T^{-2}(C_A^{-1} + C_B^{-1})(dU_A)^2$$

and since $C > 0$, $d^2S^* < 0$.
10.3 $F^* = F_A + F_B$, so that $dF^* = (\partial F_A/\partial V_A)dV_A + (\partial F_B/\partial V_B)dV_B = -P_A\, dV_A - P_B\, dV_B = (P_A - P_B)dV_B$, since $dV_A + dV_B = dV = 0$. Thus, not surprisingly, $P_A = P_B$ in equilibrium. Again

$$\begin{aligned} 2d^2F^* &= (\partial^2 F_A/\partial V_A^2)dV_A^2 + (\partial^2 F_B/\partial V_B^2)dV_B^2 \\ &= -[(\partial P_A/\partial V_A) + (\partial P_B/\partial V_B)]dV_B^2 > 0, \end{aligned}$$

i.e. the fluid must have an equation of state such that $\partial P/\partial V < 0$ which is again not very surprising. (Actually, more weakly, it would suffice that $\partial^s P/\partial V^s = 0$ $(s = 1, \ldots, r - 1)$, $\partial^r P/\partial V^r < 0$ for the particular state V, T, where s is some positive integer.)

10.4 $\partial G/\partial n_j = \mu_j = \partial(\Sigma n_i \mu_i)/\partial n_j = \mu_j + \Sigma n_i (\partial \mu_i/\partial n_j)$, whence $\Sigma n_i g_{ij} = 0$ by inspection. The condition $dG^* = 0$ of course gives $\Sigma \mu_i v_i = 0$ as usual. Next,

$$\begin{aligned} d^2G^* &= \tfrac{1}{2} \Sigma \Sigma g_{ij} \, dn_i \, dn_j = \tfrac{1}{2} (\Sigma \Sigma g_{ij} v_i v_j) d\xi^2 \\ &= -\tfrac{1}{4} \Sigma \Sigma g_{ij} n_i n_j (v_i/n_i - v_j/n_j)^2 \, d\xi^2, \end{aligned}$$

since the additional terms vanish in view of the identity already verified. Hence if $g_{ij} < 0$ $(i \neq j)$ $d^2G^* > 0$.

The condition is sufficient but it is not necessary: we have here not used the fact that $\Sigma \mu_j v_j = 0$ at all, and it may well be possible for $\Sigma \Sigma g_{ij} v_i v_j$ to be positive as a consequence of $\Sigma \mu_j v_j$ being zero even though the g_{ij} $(i \neq j)$ are not all negative.

10.5 Since P is fixed there is only the one external variable T: $n = 1$; there are two constituents, i.e. water and salt: $z = 2$; and there are three phases, i.e. the liquid solution, water vapour and solid salt: $\varphi = 3$. There is no chemical reaction, so that $R = 0$. It follows that $f = 1 - 3 + 2 - 0 = 0$, which implies that the values of the intensive variables c and T must remain fixed.

11.1 By definition $C = TS_T(V, T)$, $C^* = TS_T(P, T) = T[S_T(V, T) + S_V(V, T)V_T]$. Hence $C^* - C = TS_V(V, T)V_T$ and the required result is obtained by using the second Maxwell relation. For the result in terms of bulk coefficients see *Note* 36.

11.2

$dG = d(U - TS + PV) = dU - TdS - SdT + PdV + VdP = VdP - SdT + \Sigma \mu_j \, dn_j.$

The required relations follow by inspection, granted that $(y_1, y_2, \ldots) = (P, T, n_1, n_2, \ldots)$.

11.3 $\triangle c^* = T\partial(\triangle s)/\partial T$ by definition, so that, on taking the temperature derivative to correspond to the maintenance of equilibrium,

$$\begin{aligned} \triangle c^* &= T\{d(\triangle s)/dT - [\partial(\triangle s)/\partial P] dP/dT\} \\ &= T\{d(\lambda/T)/dT + [\partial(\triangle \mathrm{v})/\partial T]\lambda/T\triangle \mathrm{v}\}, \end{aligned}$$

in view of the third Maxwell relation and the Clapeyron-Clausius equation. Thus

$$\triangle c^* = Td(\lambda/T)/dT + \lambda \, \partial(\ln \mathrm{v})/\partial T.$$

11.4 In the Clapeyron-Clausius equation take λ constant and $\triangle \mathrm{v} = \mathrm{v} =$ the molar volume of the vapour. Then, since $\mathrm{v} = RT_\varphi/P$, we have $dP/dT_\varphi = \lambda P/RT_\varphi^2$, whence the required result follows.

11.5 It doesn't. $v_1 = -1$, $v_2 = -1$, $v_3 = 2$, so that $v = 0$.

12.1 We have $U_V = TP_T - T = \gamma T^2 V^{-\frac{1}{2}}$ and therefore $U = 2\gamma T^2 V^{\frac{1}{2}}$ + a function of T only which must vanish if U_T is to be proportional to $V^{\frac{1}{2}}$. Also $V_P = 1/P_V = -2V^{\frac{3}{2}}/\gamma T^2$. Each of the terms in question now becomes $2\gamma k T^3 V^{\frac{1}{2}} = kTU$.

12.2 $S^\dagger$ now involves also the additional term $-\mu^\dagger dn^\dagger/T^\dagger$ so that $\sigma = -\triangle S + (\triangle U + P\triangle V - \mu\triangle n)/T$.

12.3 Since $dS = (dU + PdV - \mu dn)/T$, $TS_T - U_T = 0$ as before and $TS_n - U_n = -\mu$. Therefore $a_{12} = 0$ so that $\triangle_{22} = 1/2a_{22}$, and $2kTa_{22} = -T\partial^2 S/\partial n^2 + \partial^2 U/\partial n^2 = \partial \mu/\partial n$.

12.4 In the eleventh lecture we found that (for one gas) $\mu = g = RT \ln P + \chi(T)$. Here we must use V in place of P, so that $\mu = -RT \ln (V/n) +$ a function of T. Then $\partial\mu/\partial n = RT/n$, from which the required result follows.

12.5 First, if T and n can fluctuate we proceed exactly as in Problems **12.3** and **12.4**, with the result that $\triangle'_{11} = 1/2a_{11} = kC^{-1}T^2$, $\triangle'_{12} = 0$, $\triangle'_{22} = kn/R$. Now, $\triangle U = C\triangle T + (U/n)\triangle n + \ldots$, bearing in mind that $U = n$ times a function of T only. Hence $\triangle_{11} = (\delta U)^2 = C^2\triangle'_{11} + (U^2/n^2)\triangle'_{22}$, and, since $U = CT$, the required result follows immediately.

13.1 (i) There are N molecules with $i = 2$ when $s = 2$ and $i = 3$ when $s \geqslant 3$. Therefore $r = 5N$ with $s = 2$ and $r = 6N$ when $s \geqslant 3$. (That the atoms are structureless is intended to mean that inertially they are mass-points so that they can have no 'internal motions'.)

(ii) There are sN atoms with no geometric constraints between them and therefore $r = 3sN$.

13.2 If Φ is the electrostatic potential, the potential energy of one atom at x, y, z is $e\Phi(x, y, z)$. Its kinetic energy is $\frac{1}{2}m(\dot{x}^2 + \dot{y}^2 + \dot{z}^2)$, and therefore that of K as a whole is $\sum_{k=1}^{3N} \frac{1}{2}m\dot{q}_k^2$, where the cartesian coordinates of the jth atom are $q_{3j-2}, q_{3j-1}, q_{3j}$. Also $p_k = m\dot{q}_k$. Then for values of the q_k corresponding to the interior of the enclosure,

$$H = (1/2m) \sum_{k=1}^{3N} p_k^2 + e \sum_{j=1}^{N} \Phi(q_{3j-2}, q_{3j-1}, q_{3j}).$$

13.3 We have $\varphi_0 \int d\Gamma = 1$. Integrating first over the q_k, we get a factor V^N since there is no contribution to the integral for values of the q_k which do not correspond to the interior of the enclosure. The integral over the p_k extends over the interior of the $3N$-sphere $\Sigma\, p_k{}^2 = 2mE\ (= R^2$, say). From the given result its volume is $[\pi^{\frac{1}{2}n}/(\frac{1}{2}n)!]R^n$, $(n = 3N)$. It follows that $\varphi_0 = (\frac{3}{2}N)![(2\pi mh^{-2}E)^{\frac{3}{2}}V]^{-N}$.

13.4 $U = \int H\varphi\, d\Gamma = \varphi_0\, (h^{-3}\, V)^N\, [n\pi^{\frac{1}{2}n}/(\frac{1}{2}n)!] \int_0^R (r^2/2m)r^{n-1}dr$. Using the result of the preceding problem it follows that $U = E/(1 + 2/3N)$. Since $2/3N$ is quite negligible compared with unity, we therefore have $U = E$.

13.5 $S = -k \int \varphi_0 \ln \varphi_0\, d\Gamma$, so that, since $\varphi_0 \int d\Gamma = 1$, $S = -k \ln \varphi_0$. Hence $S = Nk(\frac{3}{2} \ln U + \ln V) +$ constant, whence the required result follows at once.

13.6 We can start with canonical variables p^* and q^* for which $d\Gamma = h^{-n}\, dp_1^*\, dq_1^* \ldots dq_n^*$, the choice of the q_k^* being freely at our disposal. We therefore contemplate the transformation $q_k^* = q_k$, $p_k^* = f_k\, (\dot{q}, q) \equiv \partial L(\dot{q}, q)/\partial\dot{q}_k$. Then $\partial q_k^*/\partial q_l = 1$ or 0 according as $k = l$ or $k \neq l$, $\partial q_k^*/\partial\dot{q}_l = 0$, $\partial p_k^*/\partial\dot{q}_l = \partial^2 L/\partial\dot{q}_k\partial\dot{q}_l$, $\partial p_k^*/\partial q_l = \partial^2 L/\partial\dot{q}_k\partial q_l$. The Jacobian of the transformation is then just the determinant D of the $\partial^2 L/\partial\dot{q}_k\partial\dot{q}_l$. Therefore

$$d\Gamma = h^{-n}\, D\, d\dot{q}_1\, dq_1\, d\dot{q}_2 \ldots dq_n.$$

14.1 The answer usually given is that H is simply the kinetic energy, i.e. $H = \sum_{j=1}^{3N} p_j^2/2m$. Strictly speaking, however, one should add to this a term $\theta(q)$ where $\theta(q) = 0$ or ∞ according as the values of the q_k correspond to points inside or outside the box. $\theta(q)$ is therefore the potential of an external force. (Note that we have assumed the motions to be non-relativistic; cf. Problem **14.6**).

14.2 $Z = a \int e^{-H/kT}\, d\Gamma = a\, h^{-3N} \int \exp[-(\sum_{j=1}^{3N} p_j^2/2m + \theta)/kT]\, dp\, dq$, ($dp = dp_1\, dp_2 \ldots dp_{3N}$, etc.). Integrating over the $3N$ (cartesian) position coordinates of the particles $\int e^{-\theta/kT} dq = (\triangle q_1)(\triangle q_2) \ldots (\triangle q_{3N}) = V^N$, where $\triangle q_k$ is the range of q_k corresponding to the interior of the box. The other factor of Z involves the product of $3N$ like integrals of the form $\int_{-\infty}^{\infty} \exp(-p_j^2/2mkT) dp_j = (2\pi mkT)^{\frac{1}{2}}$, and the stated result so follows. Then $F = -NkT(\frac{3}{2} \ln T + \ln V + \text{constant})$, so that $P = -\partial F/\partial V = NkT/V$, $U = \frac{3}{2}NkT$, $C = \frac{3}{2}Nk$. (The second equation is consistent with the Theorem of Equipartition.) We are obviously concerned with an ideal gas whose specific heat is constant.

14.3 Purely classically all magnetic effects are the result of interactions between electromagnetic fields and charged particles in motion. Given a system K in an external magnetostatic field whose potential is $\mathbf{A}$ the Hamiltonian of K therefore consists additively of terms of the form $|\mathbf{p} - e\mathbf{A}/c|^2/2m$. Here it suffices to use cartesian coordinates. In the course of evaluating Z we must integrate over all the p_k from $-\infty$ to $+\infty$, and the presence of the magnetic potential in the exponent represents only an irrelevant shift of origin of each p_k. Hence F is independent of the field and so the magnetic moment of K must vanish.

14.4 (i) Let the internal motion of each particle with respect to its centre of mass be described in terms of polar coordinates α, β so orientated that α is the angle between the magnetic moment and $\mathbf{B}$. Then each particle contributes a term $-\mu B \cos \alpha$ to H over and above its contribution when $B = 0$. Therefore Z contains an additional factor $\int_0^{2\pi} \int_0^{\pi} \sin\theta \times \exp(y \cos\theta)\, d\theta\, d\varphi$ for each particle, and the required result follows immediately.

(ii) This follows by using the equation $U = kT^2\ \partial \ln Z/\partial T$, and omitting the contribution by Z_0 to U.

(iii) Since $U_m = -MB$, where M is the magnetic moment of K in the direction of the field, $M = N\mu(\coth y - y^{-1})$, so that, when $y \ll 1$, $M = \frac{1}{3}N\mu y = (N\mu^2/3kT)B \equiv \chi B$. ($\chi T$ is known as *Curie's constant*.)

14.5 Yes. We again have $Z = V^N \times$ function of T only, so that $P = kT\partial \ln Z/\partial V = NkT/V$ as before.

14.6 In X_{ij} take $i = j$ and χ_i to be any component of momentum, p, so that $\overline{p\ \partial H/\partial p} = kT$. Now $H = \sum_{s=1}^{N} [c(m^2c^2 + p_1^2 + p_2^2 + p_3^2)^{\frac{1}{2}}]_s = \sum h_s$, say, so that $\overline{c^2p^2/h_s} = kT$, where p is of course one of the three components of momentum which enter into h_s. Summing over these components, $3kT = \overline{(h_s^2 - m^2c^4)/h_s}$. If ε_s is the energy of the sth particle, excluding its rest energy, $\overline{\varepsilon_s(\varepsilon_s + 2mc^2)/(\varepsilon_s + mc^2)} = 3kT$. Since $U = \sum \bar{\varepsilon}_s$, the required results now follow by inspection on letting mc^2 formally go to infinity and zero respectively.

14.7 The conditions under which the theorem is valid are not satisfied, since $H \to -\infty$ as $q \to -\infty$, and Z does not converge. (Correspondingly, in quantum mechanics such an oscillator has no bound states.)

15.1 $H_{N_1N_2} = \sum_{j=1}^{N_1} p_j^2/2m_1 + \sum_{j=1}^{N_2} p_j^2/2m_2$. Therefore, since the integral for $Z_{N_1N_2}$ splits up into the product of $N_1 + N_2$ integrals as usual (apart from a factor $V^{N_1+N_2}$), $Z_{N_1N_2} = \tau_1^{N_1} \tau_2^{N_2} V^{N_1+N_2}/(N_1!N_2!)$, with $\tau_i = (2\pi h^{-2} m_i kT)^{\frac{3}{2}} = \kappa_i T^{\frac{3}{2}}$, say. Therefore, at once, $Z = \exp[(\lambda_1\tau_1 + \lambda_2\tau_2)V]$, so that $X = kT \ln Z = k(\kappa_1\lambda_1 + \kappa_2\lambda_2)T^{\frac{5}{2}}V$. It thus emerges on carrying out the usual differentiation that S is simply the sum of the entropies of the constituents as if each were present in the box by itself, i.e. $S = S_1 + S_2$, where $S_i = \bar{N}_i k[\ln(\kappa_i T^{\frac{3}{2}}) + \ln(V/\bar{N}_i) + \frac{5}{2}]$.

15.2 (i) Here $S = S_1 + S_2$ where S_i is as in the preceding problem, except that, of course, V has to be replaced by $V_i = \bar{N}_i kT/P$. (ii) After diffusion the entropy is given by the answer to the preceding problem. Writing $V = (\bar{N}_1 + \bar{N}_2)kT/P$ in this and mutually subtracting the two values of $S_1 + S_2$, the given result emerges at once. (iii) (The apparent contradiction is usually known as *Gibbs' Paradox*.) There is no paradox: when the two gases are identical the calculations in the preceding problem are meaningless, since N_1 and N_2 have no separate existence. In short, particles either are different from each other or they are not—there is no continuous variable which can measure a 'degree of difference'.

15.3 (i) We have $f(x) = f_0 + f_0'\xi + \frac{1}{2}f''_0\xi^2 + \ldots$, where $f_0 = f(x_0)$, $f_0' = df(x)/dx$ at $x = x_0$, etc., and $\xi = x - x_0$. Since $f(x)$ is stationary at $x = x_0$, $f_0' = 0$ and so $J = e^{-f_0}\int_0^\infty \exp(-\frac{1}{2}f_0\xi^2 - \ldots)d\xi$. By hypothesis the main contribution to the integral comes from the immediate vicinity of the point $x = x_0$ and we may so approximate J by the expression $e^{-f_0}\int_0^\infty \exp(-\frac{1}{2}f''_0\xi^2)d\xi$. Therefore $J = (2\pi/f''_0)^{\frac{1}{2}}\,e^{-f_0}$.

(ii) In the case of $N!$, $f(x) = x - N\ln x$, $f' = 1 - N/x$, $f'' = N/x^2$, so that $x_0 = N$, $f''_0 = 1/N$. Hence $N! \sim (2\pi)^{\frac{1}{2}}\,N^{N+\frac{1}{2}}e^{-N}$.

15.4 By inspection of Z, $\partial Z/\partial P = -\bar{V}Z/kT$ and $\partial^2 Z/\partial P^2 = \overline{V^2}Z/k^2T^2$. Therefore $(\delta V)^2 = \overline{V^2} - \bar{V}^2 = k^2T^2\partial^2\ln Z/\partial P^2$. Since $\bar{V} = \partial G/\partial P = -kT\partial\ln Z/\partial P$, the given result follows directly. Note that for an ideal gas $\partial\bar{V}/\partial P = -NkT/P^2$, so that $\delta V/\bar{V} = \bar{N}^{-\frac{1}{2}}$ exactly.

15.5 We take it for granted that the sum is closely approximated by the integral $J = \int_0^\infty g(N)\,dN$. Then, setting $f(N) = -\ln g(N)$ in Problem **15.3**, $J = (-2\pi/(\ln g)''_0)^{\frac{1}{2}}g_0$. Here $g = y^N/N! \sim (2\pi)^{-\frac{1}{2}}\,N^{-N-\frac{1}{2}}(ey)^N$, so that $(\ln g)''_0 \sim -1/N_0$ and therefore $J \sim (2\pi N_0)^{\frac{1}{2}}$ times the largest term of the series.

15.6 No. It only generates an additive term $-k\ln V_0$ to the entropy. Being independent of all thermodynamic coordinates its presence is irrelevant, i.e. it has no observable consequences.

16.1 It will suffice to consider the case of a single constituent. Then $X^* = G - F = \Sigma\int[\mu N - H_N - kT\ln(N!\varphi_N^*)]\varphi_N^* d\Gamma_N$, and if $\varphi_N^* = \varphi_N\exp\eta_N$, we find, since $\ln(N!\varphi_N) = (\mu N - X - H_N)/kT$, that

$$\triangle X = -kT\Sigma\int\eta_N\exp\eta_N\,d\Gamma_N = -kT\Sigma\int[(\eta_N - 1)\exp\eta_N + 1]d\Gamma_N.$$

The expression in square brackets is non-negative for every value of N, so that $\triangle X < 0$ unless $\eta_N = 0$ (all N).

16.2 If the ratio ν_1/ν_2 is an irrational number all weights are unity. When it is not $\nu_1/\nu_2 = r/s$, where r, s are positive numbers (without common divisor). Then any particular level is $\varepsilon = \frac{1}{2}h(\nu_1 + \nu_2) + (h\nu_2/s)n$, where n is an integer of the form $rn_1 + sn_2$. Hence the required weight of the level with given n is the number of solutions in integers $\geqslant 0$ of the equation $rn_1 + sn_2 = n$. The isotropic case $\nu_1 = \nu_2$ is of especial interest. Its weight is then obviously $n + 1$.

16.3 In the non-relativistic situation we set $\varepsilon = |\mathbf{p}|^2/2m$, where $\mathbf{p}$ is the momentum of a particle. Relativistically, however, $\varepsilon = c(m^2c^2 + |\mathbf{p}|^2)^{\frac{1}{2}} - mc^2$, where m is now the rest mass and c the speed of light, the rest energy not being included in ε. Thus $g(\varepsilon)$ may be obtained from its non-relativistic counterpart simply through the formal substitution

$\varepsilon \to [(\varepsilon + mc^2)^2/c^2 - m^2c^2]/2m = \varepsilon + \varepsilon^2/2mc^2$. This immediately gives the quoted result. [According to *Note* 60 the momentum components must be integral multiples of $h/2l$. This result holds equally in the relativistic domain.]

17.1 Here $H = \Sigma(p^2 + \kappa^2 q^2)/2m$, with $\kappa = 2\pi\nu m$. Therefore $Z = z^N$, where $z = h^{-1}\int_{-\infty}^{\infty}\int_{-\infty}^{\infty} \exp[-(p^2 + \kappa^2 q^2)/2mkT]dpdq = 2\pi mkT/h\kappa = kT/h\nu\,(= 1/\zeta$, say). Next $F = NkT \ln \zeta$, so that $S = Nk[1 + \ln(1/\zeta)]$, $U = NkT$ and $C = Nk$. The last result also follows directly from the Theorem of Equipartition. (There is no d-coordinate, unless one rather artificially thinks of the binding constant κ as being under one's control. However, $\partial U/\partial\kappa = 0$.)

17.2 The levels of the oscillator are $\varepsilon_j = (j + \frac{1}{2})h\nu$. Hence $z = \sum_{j=0}^{\infty} e^{-(j+\frac{1}{2})\zeta} = e^{-\frac{1}{2}\zeta}/(1 - e^{-\zeta})$, so that $F = N[\frac{1}{2}h\nu + kT\ln(1 - e^{-\zeta})]$. In turn $S = Nk[\zeta/(e^{\zeta} - 1) - \ln(1 - e^{-\zeta})]$, $U = Nh\nu[\frac{1}{2} + 1/(e^{\zeta} - 1)]$, $C = Nk\zeta^2 e^{\zeta}/(e^{\zeta} - 1)^2$.

17.3 We distinguish the quantities calculated classically by a prime. Then (i) as $T \to 0$, $S' \sim Nk \ln T \to -\infty$, in conflict with the Third Law. On the other hand, $\zeta \to \infty$, so that $S \sim Nk\zeta e^{-\zeta} \to 0$, as $T \to 0$. (ii) An equal amount of $\frac{1}{2}kT$ is contributed to U' by each degree of freedom. However, this equipartition fails in the quantum mechanical case, since each degree of freedom contributes to U an amount which exceeds $\frac{1}{2}kT$ by $\frac{1}{2}h\nu[\frac{1}{2} + 1/(e^{\zeta} - 1) - 1/\zeta]$. (iii) C' = constant, whereas it should go to zero with T. C on the other hand goes to zero with T very rapidly since, as $\zeta \to \infty$, $C \sim Nk\zeta^2 e^{-\zeta}$.

17.4 Here we have to evaluate $z = \sum_{m=-J}^{J} e^{\zeta m}$, where $\zeta = \kappa B/kT$. The sum is elementary: $[\sinh(J + \frac{1}{2})\zeta]/\sinh(\frac{1}{2}\zeta)$ and this replaces the previous function $\sinh y/y$. When ζ, i.e. B/T, is sufficiently small, $z \sim (2J + 1)[1 + \frac{1}{6}J(J + 1)\zeta^2]$. (Arguing as in the classical case, we infer from this that $\chi = J(J + 1)N\kappa^2/3kT$.)

18.1 A molecule, treated as a rigid body, has 0, 2 or 3 rotational degrees of freedom, depending on whether it is effectively 1-, 2- or 3-dimensional. Hence, in units of Nk, the corresponding rotational specific heats are 0, 1, $\frac{3}{2}$ and $C = \frac{3}{2}, \frac{5}{2}$, 3 respectively. (We are clearly concerned with classical conditions.) Also, the gases being nearly enough ideal, $C^* - C = 1$, and therefore $\gamma = 1 + 1/C = \frac{5}{3}, \frac{7}{5}, \frac{4}{3}$, respectively. Argon and oxygen are therefore adequately accounted for. Not so ethylene, for which $\gamma = \frac{5}{4}$, corresponding to $C = 4$. The excess is contributed in this case by internal vibrations.

18.2 The question is not well-defined because we are not told anything about the character of the interatomic forces responsible for vibrations. If the atoms are structureless and the temperature low enough, the vibrations will be harmonic and by the Theorem of Equipartition, $C = (3s - \frac{5}{2})Nk$ if the molecule is linear and $C = (3s - 3)Nk$ when it is not. Therefore $\gamma = (6s - 3)/(6s - 5)$ and $\gamma = (3s - 2)/(3s - 3)$, respectively.

18.3 Here $\rho = 1$ and $\tau = 1$ so that the electronic specific heat is $e(1 + e)^{-2} \approx 0.196$. The vibrational specific heat is zero. Hence the predicted value of $C \approx 2.696$, which agrees well enough with experiment.

18.4 $\rho = 3\zeta_-/\zeta_+$, so that we require the root of the equation $9\xi(1 + \frac{7}{3}\xi^5 + \ldots) = 1 + 5\xi^3 + \ldots$, where $\xi = e^{-2\tau}$. $\xi = 1/9$ is a first approximation to the root, $\xi \approx 0.1104$ a better one. This gives $T_{\rho=1} \approx 0.907\ T_r$.

18.5 Excitation energies are measured on a scale of millions of electron volts $\approx 1.6 \times 10^{-6}$ erg. The corresponding characteristic temperature is, nearly enough, 10^{10} °K. At such temperatures matter is completely ionized, relativistic effects must be taken into account, including electron-positron creation, and so on.

19.1 $u = \int_0^\infty u(\nu)d\nu = 8\pi c^{-3}h\int_0^\infty \nu^3 d\nu/(e^{h\nu/kT} - 1) = 8\pi k^4 c^{-3} h^{-3} T^4 \int_0^\infty x^3\,dx/(e^x - 1)$. The integral has the value $\pi^4/15$. Hence $\sigma = 8\pi^5 k^4/15c^3h^3$.
19.2 As far as 'mechanical quantities' such as P and U are concerned, all thermal motions are irrelevant at $T = 0$—the system is effectively purely mechanical. Therefore the statistical constant k cannot appear in the relation between P and V.
19.3 All the lowest energy levels are filled, up to some energy ε_1, say. All others are empty. Therefore ε_1 is determined by the condition $\int_0^{\varepsilon_1} g(\varepsilon)d\varepsilon = N$, and then $U = \frac{3}{2}PV = \int_0^{\varepsilon_1} \varepsilon g(\varepsilon)d\varepsilon$. With $g(\varepsilon) = A\varepsilon^{\frac{1}{2}}$, where $A = 2\pi\omega h^{-3}(2m)^{\frac{3}{2}}V$, we thus have $U = \frac{2}{5}A\varepsilon_1^{\frac{5}{2}}$, $N = \frac{2}{3}A\varepsilon_1^{\frac{3}{2}}$. Eliminating ε_1, we get the previous result $PV^{\frac{5}{3}} = (6/\pi\omega)^{\frac{2}{3}} N^{\frac{5}{3}}(h^2/20m)$. Note that ε_1 is the same as ε_F.
19.4 As in the preceding problem, we have $N = \int_0^{\varepsilon_F} g(\varepsilon)d\varepsilon$. Using the result of Problem **16.3**, we thus have the equation $N = 4\pi\omega(ch)^{-3}\,V\int_0^{\varepsilon_F} (\varepsilon^2 + 2mc^2\varepsilon)^{\frac{1}{2}}(\varepsilon + mc^2)d\varepsilon$ for ε_F. The integral has the value $\frac{1}{3}(\varepsilon_F{}^2 + 2mc^2\varepsilon_F)^{\frac{3}{2}}$, so that $\varepsilon_F = mc^2[(1 + \Gamma)^{\frac{1}{2}} - 1]$, where $\Gamma = (3h^3N/4\pi\omega m^3c^3V)^{\frac{2}{3}}$. In particular, when $N/V \to 0$, $\Gamma \to 0$ and we are immediately led back to the previous result. On the other hand, when $N/V \to \infty$, $\varepsilon_F \to mc^2\Gamma^{\frac{1}{2}}$, i.e. $\varepsilon_F = hc(3N/4\pi\omega V)^{\frac{1}{3}}$, which is independent of m, as we would expect.
19.5 We have quite generally $X = kT\int_0^\infty g(\varepsilon)\ln(1 + \lambda\,e^{-\beta\varepsilon})d\varepsilon$. Hence $\overline{N} = \partial X/\partial\mu = \beta\lambda\partial X/\partial\lambda = \int_0^\infty g(\varepsilon)d\varepsilon/(\lambda^{-1}e^{\beta\varepsilon} + 1)$. In the limit $T \to 0$ (which defines ε_F) $\overline{N} = \int_0^{\ln\lambda/\beta} g(\varepsilon)d\varepsilon = \int_0^{\varepsilon_F} g(\varepsilon)d\varepsilon$. On the other hand, when $T = T_0$ the chemical potential vanishes, so that $\lambda = 1$. Therefore $\overline{N} = \int_0^\infty g(\varepsilon)d\varepsilon/(e^{\beta_0\varepsilon} + 1)$. The required relation is thus $\int_0^{\varepsilon_F} g(\varepsilon)d\varepsilon = \int_0^\infty g(\varepsilon)d\varepsilon/(e^{\varepsilon/kT_0} + 1)$. When $g(\varepsilon) = A\varepsilon^r$ this becomes $(5T_F/2T_0)^{r+1} = (r + 1)\int_0^\infty t^r dt/e^t + 1)$.

20.1 It is only necessary to put $g = 2$ in the results worked out for general j. Then straight away

$$S = Nk[\ln(2\cosh\xi) - \xi\tanh\xi], \qquad U = U_0\tanh\xi, \qquad C = Nk\xi^2\,\mathrm{sech}^2\,\xi.$$

20.2 Referring to the preceding problem, C has an extremum when ξ sech ξ has, that is when $\xi\tanh\xi = 1$. Therefore C is a maximum when $\xi \approx \pm 1.200$, and then $C/Nk \approx 0.439$.
20.3 Yes. This would be the case if the energy had neither an upper nor a lower bound. (It is not irrelevant to recall that in Maxwellian (non-quantal) electrodynamics the energy of a pair of oppositely charged point particles is not bounded in either direction.)
20.4 Yes. When $T < 0$ the entropy is a decreasing function of the energy, so that this time at the end of a number of cycles S will have increased if U has decreased. The previous remark that U has a lower bound is very relevant here.
20.5 (i) Yes. (ii) No. In an adiabatic quasi-static process the entropy is constant, and so therefore is B/T. Hence when $T < 0$ a decrease of B implies an increase of T, i.e. the system heats up.

LIST OF PRINCIPAL SYMBOLS

Roman Symbols*

a_N	a_N^*/a_N, 52
B	magnitude of magnetic induction, 50
B_r	r'th virial coefficient, 37
c	speed of light, 50
c_j	concentration of j'th constituent, 25
C	specific heat at constant volume, 28; or at constant d-coordinates, 81
°C	specifies Celsius scale of temperature, 18
C_j	j'th constituent, 23
C^*	specific heat at constant pressure, 38; or under general conditions, 81
E_j, E_{Nj}	energy of j'th stationary state, 57, 59
f	number of (thermodynamic) degrees of freedom, 24
f_{ij}	$\exp(-\beta u_{ij}) - 1$, 62
F	Helmholtz potential, 24
g	molar Gibbs function, 37
g_j	weight of level E_j, 57
$g(\varepsilon)$	density of states, 57
G	Gibbs function, 24
h	Planck's constant, (45), 60
H	Hamiltonian function or operator, 44, 58
H_N	as H, with specified particle number, 51, 59
i	number of internal degress of freedom, 44
k	Boltzmann's constant, (40), 48, 53
K	generic symbol for given system, 7
K_0	as K, under condition of adiabatic isolation, 8
°K	specifies absolute (Kelvin) scale of temperature, 28
$\mathscr{K}$	generic symbol for representative ensemble of K, 40
K_c	equilibrium constant, 37
l	thermal wavelength $[=(2\pi mh^{-2}kT)^{-\frac{1}{2}}]$, 62
L	Avogadro's number, 51
m	mass of particle, 46
m	magnetic quantum number, 61
n	number of (external) coordinates, 5, 23
$\bar{n}$	Σn_i, 25
n_j, n_{aj}	number of moles of j'th constituent (in a'th phase), 23
n_{ji}	number of particles in ith one-particle level contributing to E_{Nj}, 59
N	number of particles in closed system, 44
$\bar{N}$	(mean) number of particles in open system, 51
N_j	number of particles of C_j, 51
O	if $X=\mathrm{O}(x^n)$ in some limit, X/x^n tends to a non-zero constant in this limit, 68
p	the set $(p_1, p_2, \ldots p_r)$, 44
p_k	k'th canonical momentum, 44
P	pressure, 5
P_k	k'th generalized force, 5
q	the set $(q_1, q_2, \ldots q_r)$, 44
q_k	k'th canonical coordinate, 44
Q	heat, 14
r	number of (mechanical) degrees of freedom, 44

*The page numbers refer to the place where a symbol with the given meaning appears for the first time.

r_{ij}	mutual distance between ith and jth particles, 62
R	ideal gas constant 22, (36)
s	empirical entropy, 11
s	spin, 57
S	metrical entropy, 21
$\mathfrak{S}$	state, 7
t	empirical temperature, 17
T	absolute temperature, 21
T_F	$2\varepsilon_F/5k$, 67
u_{ij}	potential energy of mutual interaction of pairs of particles, 62
U	energy, 14
v	molar volume, 36
V	volume, 4
w	defined by $S=\overline{w}$, 47, 51
w	$\int_0^\infty [1-\exp(-\beta u)]r^2dr$, 63
W	work done by system, 5
x_k	k'th (thermodynamic) coordinate, 4
X	grand potential, 24
z	number of phases, 23
z	one-particle partition function, 60
Z	canonical partition-function, 48, 58
Z	grand canonical partition function, 52, 59
Z_N	canonical partition function, 52, 60

Greek Symbols

α	isobaric coefficient of expansion [$= \partial \ln V(P, T)/\partial T$], 35
a, a_N	multiplicative constant in canonical partition function, 48, 52
a_N^*	multiplicative constant in grand canonical partition function, 52
β	isometric pressure coefficient [$= \partial \ln P(V, T)/\partial T$], 35
β	$1/kT$, 49
Γ, Γ_N	phase space, 45, 51
$d\Gamma$, $d\Gamma_N$	element of extension in phase [$= h^{-r}dpdq$], 45, 51
δ	for any X, δX is an infinitesimal variation of X, 5
$\triangle$	for any X, $\triangle X$ is a finite change of X, 14
ε	one-particle energy, 57
ε_i	energy level of one-particle system, 57
ε_F	Fermi energy, 67
ζ	$\overline{N}l^3/\omega V$, 66
η_i	$\exp[\beta(\mu-\varepsilon_i)]$, 66
κ	isothermal compressibility [$= -\partial \ln V(P, T)/\partial P$], 35
λ	latent heat of phase change, 36
λ	absolute activity ($= \exp(\mu/kT)$], 52
μ	magnitude of magnetic moment, 50
μ_j	chemical potential of jth constituent, 23
ν	frequency of oscillator, 64
ν_j	stoichiometric coefficient, 24
ξ	degree of advancement, 24
$\xi(\lambda)$	$\frac{4}{3}\pi^{-\frac{1}{2}} \int_0^\infty [t^{\frac{3}{2}}/(\lambda^{-1}e^t \pm 1)\,dt$, 67, 68
Π	product sign, 37
σ	entropy difference between state and corresponding pseudo-state, 40
Σ	summation sign, 19
τ	$(2\pi mh^{-2}kT)^{\frac{3}{2}}$ [$= l^{-3}$], 53
τ	ratio of a characteristic to the absolute temperature, 63
φ	number of phases, 23
φ, φ_N	probability in phase, 45, 51
φ_j, φ_{Nj}	probability of state, 57, 59
ω	spin degeneracy of one-particle state, 57

INDEX